Marketing in fisheries and aquaculture

Ian Chaston BSc, PhD, MBA

Fishing News Books Ltd
Farnham, Surrey, England

British Library CIP data
Chaston, Ian
  Marketing in fisheries and aquaculture.
  1. Fisheries
  I. Title
  380. 1'437   SH211

  ISBN 0-85238-129-8

Published by
Fishing News Books Ltd
1 Long Garden Walk
Farnham, Surrey, England

Typeset by
Keystroke Ltd., Godalming, Surrey
Printed in Great Britain by
Page Bros. (Norwich) Ltd., Norwich

*In praise of the fisher of men,
and in love for Lyn, Miles and Annabel*

Contents

Figures and tables

8

Preface

The current economic problems facing the fisheries and aquaculture industry cannot be solved just by relying on Government subsidies and the continuation of quotas to stabilize total landings. These actions must be accompanied by an improvement in the business management capabilities of industry personnel and public sector advisory services, especially in the area of marketing. This is the most important business function because the purpose of marketing is to effectively promote a company's output and thereby generate sales revenue and profits. The objective of this book, therefore, is to describe how the concepts and practice of modern marketing management can be applied in the fish industry.

The early chapters focus on the philosophy and mechanism of market need identification and satisfaction. This is followed by a review of how information from this activity can provide the basis for the correct product positioning, new product opportunities and the development of appropriate marketing strategies. Execution of such strategies involves the variables of price, promotion and distribution. Planning, utilization and control of these factors are discussed in relation to specific examples drawn from various sectors of the industry.

The text assumes no prior knowledge of marketing and is designed to help existing members of the industry and public sector advisors. It will also provide the syllabus material for a college course on business studies in fisheries and aquaculture.

1
The marketing concept

A fisherman in a small rural community who lands a catch in excess of his needs and seeks to exchange the fish for another product is implicitly involved in the activity of marketing. The role of man as a 'trader' negotiating the exchange of surplus production for another resource has been a behaviour component since virtually the beginning of time. Hence it can be said of marketing that it is an ancient tradition and, in fact, is unique to humanity because no other species performs this activity.

The Industrial Revolution evolved as man began to specialize in specific economic sectors and to exploit the benefits of mass production. This was accompanied by recognition of the need to introduce analysis, planning, implementation and control into the increasingly complex world of business. From this was born the concept of formal management responsibilities within commercial organizations.

The earliest management orientation centred on the production process. Goods and services were scarce and the central problem was to find effective ways to increase output. By the end of the nineteenth century the industrial revolution had spread to most countries, and with the simultaneous improvement in transportation the world entered a new era in which industrial output exceeded demand. Revenue was no longer guaranteed just because a company was capable of producing goods.

This more competitive environment caused the larger corporations to recognize the weakness of their production orientated approach with its implicit assumption that there was always a market available for their products. Some companies decided that the starting point in the management process was to identify the requirements of potential customers. This information could then be used to develop a product capable of satisfying those needs. This has become known as the marketing orientated approach to management. It represents a total reverse in the logic path of the

11

production orientated company which places the product first and then attempts to locate a customer.

The market orientation approach requires a mechanism to identify need. This is achieved through market research to evaluate, measure and interpret the attitude and behaviour of the potential customer. A successful company uses these data as the basis for product development, with the additional objective of creating goods which offer a differential advantage over competition. Such products are then presented to the customer accompanied by a marketing programme which integrates the variables of product, price, promotion and distribution into the most effective form to ensure the establishment of an on-going exchange relationship with the customer.

The concept and the fishing industry

Despite the proven benefits of the marketing orientated philosophy of management, it remains a concept of only limited acceptance within the fishing industry. The position taken by many is that fishermen will catch a certain species depending on the season and location; competitors will be landing the same species, so that they have no control over the sale of their product because it cannot be differentiated from that of the competition. In fact one could advance the same argument for coffee beans or soap, and had it been accepted, some large multinational companies would not be the dominant forces they are now in the western hemisphere.

At the time of writing, demand and supply in the fishing industry are not in equilibrium and many companies are at the point of bankruptcy. It is not suggested that these problems (which are of course not new, they occur on a cyclical basis) could all be solved by the fishing industry accepting the marketing concept. Nevertheless it is no coincidence that the survivors after an economic downturn are usually those companies which are orientated towards identifying the changing demands of the marketplace and attempt to satisfy the buying patterns of their customers. They have recognized that one should continually monitor customer need, and furthermore, even though one fish can appear very much like another when landed, the raw material can be the basis of effective differentiation to ensure a sustained demand for particular products even when total market volume is declining.

One example of this marketing approach is provided by the Icelandic fishing industry. They start with the advantage that Icelandic fish is

1. Citibank matl's
2. BA to JLD — We call them
3. <u>P&G Confirmation</u>
4. Demid mcm:
   RGE
   a. matl's to Brity
   b. Talk to Bob White

5. AQUA
   A. Identify users

6.

among the finest in the world, an attribute maintained by attention to quality control during processing and distribution. The superiority of the product is communicated through an integrated marketing programme which has brought Icelandic fish to the point where it can command a premium price in the marketplace.

One possible factor which is often advanced to support the production orientation approach in the fishing industry is that current world catch is static and, with demand increasing, the shortage of fish resulting will mean there is always a market available for whatever is caught. This logic frequently forms the basis both of private and public sector projects in various parts of the world on the potential for the development of a new fishery to utilize underexploited species. In many instances no real attempt is made to evaluate market acceptance of the species. The participating companies merely proceed to make major investments in fishing vessels and processing facilities to exploit the new resource. Compared to the scale of costs associated with this action, a market research survey prior to such investment would represent a small increment in cost, yet yield significant information to reduce the risks of the new species not being capable of establishing itself in the marketplace against other competitive products.

In the early seventies, for example, the North American market needed a lower cost alternative to cod because of reduced availability of this species. The South American whiting resource off the coast of Argentina had been known to exist for some time. A number of companies felt that whiting could provide a substitute for cod and invested heavily in the establishment of a new fishery. Concurrent to this programme, the same reasoning had caused the Japanese and Korean trawlers to begin their fishing for another substitute species, Alaska pollack, in the northern hemisphere.

The South American whiting has a characteristic heavy fat layer below the skin which can cause rancidity in the finished product, especially if insufficient attention is paid to handling between catching and processing. Within a short time, North American manufacturers of breaded fish products began to reject South American fillets because of a rancidity problem and switched to Alaska pollack, which, in terms of other features such as texture and moisture content, had been considered inferior.

A marketing orientated approach in the South American fishery would have identified the demand for a white, bland flesh and immediately alerted producers to the requirement to solve the rancidity problem prior

13

to investment in the new fishery. In the mid-seventies, a solution was developed which entailed a much deeper cut when skinning the fillet to remove the subcutaneous fat layer. These 'deep skin' whiting fillets were immediately accepted by the market, whereupon virtually all processors switched to the species and away from the alternative, Alaska pollack.

After introducing the new product, however, the South American industry still did not totally adopt a marketing orientated approach. They failed to recognize that deep skin whiting fulfilled the need for a low cost alternative to cod. Capacity was expanded on the apparent assumption of rising demand for the product. Initially this proved to be the case and the price for whiting fillet blocks used in the manufacture of breaded fish began to approach that of cod. In 1980 overall demand for fish in the USA began to decline due to the downturn in the overall economy. Then an improvement in the availability of cod compared to the previous decade brought a narrowing in the price differential and many producers began to switch back to cod as the preferred species. Suddenly the Argentine fishery had excess supply, with few buyers willing to commit themselves to future purchases at the high prices being asked. The South American industry had again – because of their production orientated approach to management – failed to recognize that if their product was no longer perceived as a lower cost alternative, their fish would be unable to compete.

The concept and aquaculture

Aquaculture is a relatively new industry to be adopted on a wide scale, and one could assume that pioneers of the technology would have followed the modern approach to management of matching production to market need. This has not proved to be the case and the industry has suffered a number of setbacks because of insufficient attention to marketing during the species selection stage of the research programmes.

For a 'farmed' product to compete against raw material from the traditional fishing industry the aim must be to raise the product to sell at less than current market prices. Hence early research was often focused on premium price species while failing to assess the cause of the high price in the marketplace. One does not even have to embrace a market orientation concept to understand that limited supply will create high prices.

Aquaculture is capable of significantly expanding the availability of a product and the basic economic laws of supply and demand will cause prices to fall as supplies increase. One potential example of this situation is the warmwater finfish pompano where one medium-sized farm could have an annual output equal to the total landings from the US Gulf Coast fishery. Without even constructing a farm, it can be estimated that the culture costs of the species would only be 10–20% lower than the current market price. Double the supplies through aquaculture and the market price would undoubtedly trend to a point lower than the costs of culture. Yet significant efforts were expended in the 1970s in the public and private sectors on the culturing of pompano.

Another very common approach in the culture candidate selection procedure is to identify which species can tolerate the environmental conditions in the chosen geographic location and then to focus research attention on the species which technically is the easiest to raise. The northeast coast of North America provides a suitable environment for mussels, and with the years of raising this bivalve in Europe, the transfer of suitable technology is a relatively simple matter. Accordingly a number of companies decided to farm this species in Canada and the USA. What they failed to evaluate was the market acceptance for mussels, for, unlike Europe, this shellfish is relatively unknown in North America. Hence only shortly after the culture units were established, the owners were faced with the problem that insufficient demand existed for their output. This situation is gradually changing as efforts are now being made to stimulate increased usage of mussels in American restaurants. In retrospect, however, it would have been more logical to commence this latter activity by using imported mussels from Europe and to defer the building of farms until after broadscale market acceptance had been achieved.

Another production orientated approach in the species selection process is provided by the Malaysian prawn or freshwater shrimp *Macrobrachium*. Research by individuals such as Dr Fujimura in Hawaii demonstrated that this crustacean had all the characteristics of the ideal culture candidate. Maturation could be completed in captivity, the animal tolerated poor water quality conditions, it could be raised in fresh or brackish water and as an omnivore could be fed on very simple artificial diet formulations. At that time raising marine shrimp was recognized as a much more technically difficult proposition and in the early seventies a number of companies decided that *Macrobrachium* provided an excellent farmable substitute for marine shrimp.

If one examines *Macrobrachium* in terms of a marketable product using samples readily available from the traditional fishery, a number of

15

shortcomings are immediately apparent. The exoskeleton is much tougher than marine shrimp, which means processors or restaurants face a very difficult task when attempting to peel the product before use. The delicate consistency of the fish means that if cooked using the accepted procedures for marine shrimp, one is often left with a totally uneatable product. A third problem is that the product undergoes rapid deterioration in quality if frozen prior to packing. As the majority of marine shrimp is distributed in the frozen state, this would mean that unless a very specialized freezing technology is used, *Macrobrachium* could not be marketed in the frozen food segment of the shrimp industry. Even if all these problems could be overcome, the taxonomic differences between the animal and its marine counterparts are sufficient to enable companies in the shrimp industry to gain government support in some countries for legislation to specify that it could not be marketed under the generic title of 'shrimp'.

Despite all of the above, a number of major corporations invested funds on the development of *Macrobrachium* culture. It was recognized only in the late seventies that the best near-term commercial strategy would be to market the product in the whole, fresh shell-on form to restaurants located near to the farm. Clearly this limits market potential and as many of the technical problems associated with marine shrimp have now been solved, many companies are directing their attention away from *Macrobrachium* towards marine shrimp. Although the scientific community has benefited from the research effort on *Macrobrachium*, in commercial terms there has been very little return on the original investment.

It should be stressed that the marketing orientated approach to aquaculture does not merely entail identifying customer need and leaving the technologist to raise the selected species. The marketer is responsible for matching need against the resources available to the company. Hence the role of the marketer is to produce the initial need assessment, match this against the range of technically feasible culture choices, and help senior management to decide which animal represents the optimum culture candidate relative to the financial objectives of the company. The marketer should then be ready to propose an alternative strategy if research on the culturing of the species reveals unforeseen problems, *eg* excessive production costs.

2
The marketing system and the environment

The marketing system

For a permanent exchange relationship to exist between producer and customer, some form of institutional structure must be created. The components of this structure, the marketing system in the case of fishermen supplying individuals in the same village, are relatively simple (*Fig 2·1*). There is a flow of information between those involved over price and availability, and then, if there is mutual agreement over the conditions of the sale, a physical transaction flow involving the product and payment will occur.

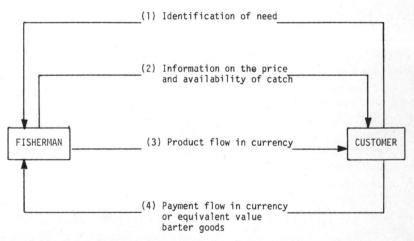

Fig 2·1   Marketing system for a producer and consumer in same location

The industrial revolution resulted in the majority of the population in many countries moving away from areas of food production into towns and cities. To supply this structure of society, the creation of a more complex marketing system using intermediaries to link the producer to the market was required (*Fig 2·2*). If a fish company was to survive, it was necessary for the management to understand the nature of this system and the variables of the marketing function which can be used to influence purchase decisions.

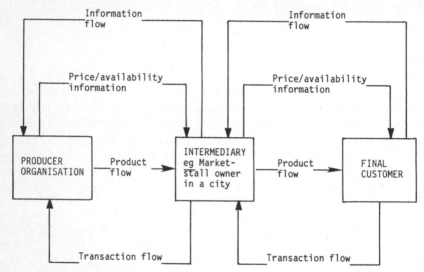

Fig 2·2   Marketing system for a final customer group located some distance from the producer

The concept can be illustrated by constructing a marketing system for a hypothetical sardine canner (*Fig 2·3*). The company has already made a 'product-market' decision to manufacture a premium quality grade sardine to be sold through grocery stores. An examination of the market will provide information on the nature of opportunities and the sources of competition.

The product is in direct competition with other brands of sardines which in turn comprise a sub-component of a larger market, the canned fish market. Even this generic product form is not the boundary of the market. Canned fish is actually embedded in a much larger entity, the food market, which represents all the items that are a source of nutrition for the population.

The objective of the marketing manager is to identify the variables

18

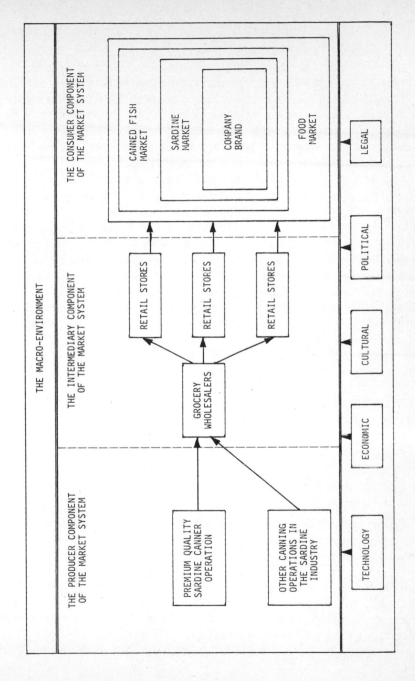

Fig 2·3   Marketing system for a premium quality sardine canner

which affect sales, and isolate those which can be influenced by the actions of the company. Total food sales, for example, tend to be a function of population size and *per capita* income. It is only as one approaches the boundary of the sub-component market for canned sardines that the company can utilize those mechanisms, or 'marketing decision variables', which cause consumers to purchase its products. Nevertheless, if the company is to be able to make predictions of future sales trends, it is necessary for management to be aware of the influence on total food sales of the non-controllable variables within the marketing system.

The nature of the variables involved in the system are described in *Fig 2.4*. Even at the sub-component market level of total canned fish, the company still has little control over consumer purchase behaviour. Cultural factors are an important influence on the consumers' willingness to accept canned fish as part of their diet. Fish is usually only one item in a range of protein foods such as meat, poultry and dairy products. Hence, in addition to the cultural issue, another more quantifiable variable is the relative price of canned fish versus other protein sources.

Once the customer has narrowed the purchase decision to canned fish, then it will require the combined effects of the marketing programmes of all the sardine canners in the market to overcome the benefits offered by other species of fish. Only at this point can the company begin to bring the marketing decision variables into play to influence directly the purchase decision.

The important variables available to the company are commonly referred to as the '4Ps of marketing mix', namely:

(*1*) Product
(*2*) Price
(*3*) Promotion
(*4*) Place (*ie* distributing the product to a location used by the customer wishing to purchase the product).

The company has previously decided that the key product benefit which will characterize the brand is premium quality. Typically this will involve the company in higher than average manufacturing costs which in turn will be reflected in an above average price in-store. The purpose of promotion will be to use the mechanisms available, *eg* advertising, personal selling, temporary value incentive such as coupons or reduced prices, to communicate information about product quality sufficient to differentiate the brand from competitors.

There would be no point in effectively applying the variables of product,

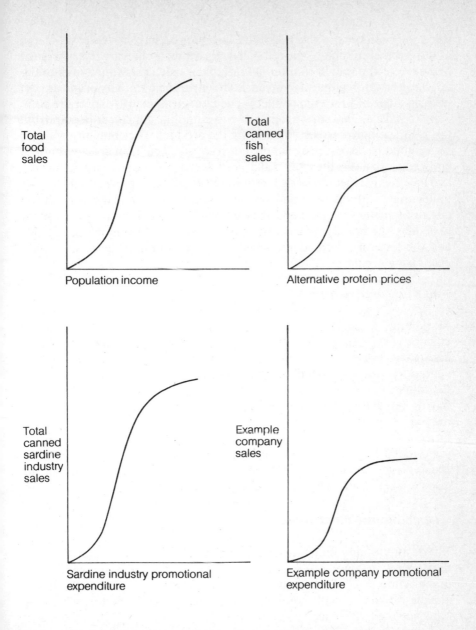

Fig 2·4   Sales and influencing variables for components of the market

21

price and promotion if the potential customers do not encounter the brand at the point of purchase, the retail grocery store. Achieving this aim is the objective of the fourth marketing variable, place. This function is orientated to persuading the intermediaries to carry the company brand as part of their range of grocery products. The intermediary will expect the company to have a marketing programme to stimulate consumer demand, but this is not sufficient to justify carrying the product. The company will also be required to have a programme covering sales calls on grocery outlets, allowances to assist the intermediaries' advertising activity, credit policies, and an effective method of delivery. In return, the company should be able to negotiate with the intermediary on product handling in the store. It will attempt to achieve agreement on issues such as the amount of shelf space to display the brand, the level of inventory to avoid running out of stock and the level of intermediary advertising to communicate in-store merchandising events.

The macro-environment

Marketing is essentially an externally focused function, and for a company to succeed it is necessary to identify the opportunities or constraints presented by changes in the environment. The most immediate contact for the organization is the market system of which it is a part. Reaction to variations in either the behaviour of the intermediaries or the buying population has already been identified as a function of management. However the total set of forces impinging on the company is much broader than just those within the market system for this system in turn is surrounded by a macro-environment which contains variables such as economic, technological, political, legal and cultural factors which are also capable of affecting the company's operations.

The economic environment

The destiny of any company is closely linked to the economic climate of the country in which it is based. In the case of the fishing industry where many companies rely heavily on export sales, management also has to include global economic conditions in any review of the factors which can influence their existence.

Trends such as higher inflation or unemployment will have a direct, adverse effect on customers' purchasing ability. Other economic variables will act on sales in a more diffuse way. The oil crisis and consequent rise in energy costs, for example, has over the last ten years significantly increased the operating costs of fishing vessels. The landed prices for fish have been rising at a rate in excess of the average global inflation rate and many species are no longer a low cost source of protein for the consumer but now represent premium priced goods which will be purchased less frequently. For the marketing manager this has required a major change in the approach used to present to the consumer the benefits of fish compared to other sources of protein.

Another diffuse economic force is the value of a country's currency relative to other nations. In recent years, the world financial markets have witnessed sudden major fluctuations in currency values as speculators have reacted to political rumours or to changes in the interest or inflation rates in various countries. If a merchant is located in a country where the currency value has appreciated relative to other nations, he may face the problem of lower cost imported fish entering the market. Also any export sales will become adversely affected because his product will now cost more on world markets. On the other hand, if the company utilizes imported fish to supply the product line, the increased value of the currency will be an advantage as it will lower the cost of raw material from overseas.

The technological environment

The most likely source of major change which can affect an industry is technological innovation. Scientific progress is therefore an issue of interest for the marketer, both from the new opportunities it can offer the operation, and, probably even more importantly, because of the potential threat posed through pre-emptive adoption by competition.

In the late sixties, the introduction into the Canadian ocean perch fishery of midwater trawling using stern trawlers resulted in a 250% improvement in catch per hour of trawler effort. For companies capable of converting their fleet to utilize the new gear technology, the benefits were an improvement in total landings and a reduction in operating costs per pound of landed fish. This was reflected by an increase in both sales revenue and profitability. However, for those operations equipped with older side trawlers, the gear conversion was not feasible. Such companies were then placed at a major disadvantage versus the competitors who could convert their boats.

23

Aquaculture as an industry owes its origins to scientific innovation by biologists and bio-engineers. Initially the concept was welcomed by the fishing industry as a potential solution to the problem of depleted natural stocks. The point has been reached where the cost of production for certain cultured species is below that of the natural fishery. This has serious implications for companies with major investments in fishing fleets. One example is the shrimp industry where 'farmed' produce is now becoming a more than minor element of world landings and in some countries, *eg* Ecuador, is reaching the point where it represents the major source of total output.

Within technologically intensive industries, the rate of innovation is often exponential. Having developed a new technology based concept, the company has to decide whether to introduce this methodology into the operation in the expectation that even this new approach will be overtaken by further innovation within a short period. The traditional fishing industry is still relatively labour-intensive and does not usually face this problem, but in aquaculture new culture techniques are announced on a regular basis. Hence in planning future programmes, the management must assume that at some indeterminable date a competitor may introduce a more cost-effective farming technique and thereby establish an advantage in the marketplace.

The above issues are concerned with technological change affecting fish production. Yet it must be recognized that one of the greatest hazards is innovation which may result in a superior substitute product being developed by organizations outside the core industry. The railroad companies certainly did not realize that Orville Wright's activities on the sands of Kittihawk represented a threat to their existence. The fishing industry therefore should accept that it is part of the food industry and carefully monitor developments such as improved vegetable proteins or advanced single cell culture techniques. One day these could represent major threats as alternative food products.

The political environment

Probably at no time in history have political and economic events been so closely linked. The majority of governments include in their political philosophy the concept of the management of economy for the benefit of the population. Although this situation may be decried by the economic disciples of Adam Smith with their belief in 'free market systems', political

24

involvement in industry is a component of the environment to be recognized as an influence on the marketing operation.

One of the earliest concerns during the industrial revolution was the ability of companies to control market trends through the creation of a monopolistic situation. Many countries now have government bodies and legislation to control monopolies or the implicit monopoly created by pricing agreements between companies within an industry. During the twentieth century, the demands of the electorate have extended into requiring protection from acts by business which are perceived as potentially detrimental to society. This protection has been established by regulations, some specific to the fishing industry, *eg* minimum quality standards, recommended quality control procedures, and others of a much broader impact across all industries through legislation on employment, working conditions, product liability and the disclosure of information.

A more recent development in the arena of political control over industry has been that governments with an ability to control the fishing industry have not been left unscathed in this situation because many countries now exert control over fish stocks through actions such as extended territorial rights over coastal waters and the licensing of foreign vessels involved in exploiting local fish stocks. These actions can cause short-term fluctuations in raw material supplies which are difficult to predict but still have to be included in management's evaluation of future market conditions. The shrimp industry, for example, has faced severe shortfalls in landings, with a consequent impact of rapidly rising prices, due to political unrest in the Middle East and Central America.

The legal environment

The exchange relationship has over the centuries been formalized through created precedents which now represent a body of legally enforceable responsibilities for the parties involved. The marketer has to understand these issues as well as continually monitor potential changes which could alter contractual terms within the marketing system. One example of this issue is that in many countries a company can no longer legally enforce price maintenance, *ie* be able to insist on specific price levels that intermediaries may charge for the company's product upon resale within the system.

Other legal forces come from the legislation created to enforce the political decisions mentioned in the previous section. Issues such as labelling, non-sustainable claims made in advertising, and the licences required to operate in specific geographic areas, are all matters which the marketer must comprehend if the company is to operate successfully under these government controls. Whether these formal mechanisms or self-regulatory actions by industry are the most effective solutions in such instances is not a debate which should really concern the marketer. It is more important that he does not ignore the legally enforceable definitions of business practice to the detriment of the company.

The cultural environment

The most diffuse component in the macro-environment is culture – that complex set of variables which forms the values and expectations of the population. It is necessary for the marketer to understand the culture of a target market because the values of the society have major influence on customer buying behaviour.

Economic growth which improves affluence is accompanied by a higher consumption of luxury goods and leisure activities. Opportunity exists in these circumstances to expand sales for more expensive products and also to anticipate a decline in demand for lower cost, more mundane, fish species. The era of greater travel and global communication has also changed the influence of parental values to the point where certain ethnic foods are no longer considered a vital component of regular diet. This situation has been used as a partial explanation for the decline in sales of smoked and certain canned fish products in North America. Previous generations brought their European eating habits to the country but gradually the more cosmopolitan modern-day society is now rejecting these items in preference for other types of fish.

The problem for the marketer is that many cultural changes are gradual. It is therefore difficult to measure a variation that could mean a decline in sales for current products or define new opportunities at an optimal point in the planning process.

3
Markets and buyer behaviour

Market definitions

As one would expect, the marketing of Peruvian anchoveta to fish meal processors requires very different management skills to those needed in selling Nova Scotian smoked salmon to housewives in Montreal. Every market exhibits unique features during the transaction process. A number of approaches have been proposed to handle these differences, but one of the simplest and most widely accepted is that of dividing markets into two basic types, the consumer market and the industrial market.

Consumer markets are concerned with products purchased by individuals and households for personal consumption. Products in the industrial market are purchased as an element in an activity to generate a tangible economic return for the buyer. In some instances this element may be a raw material which undergoes further processing by the buyer, *eg* fresh prawns purchased by a manufacturer for use in a seafood entrée. Other buyers may not modify the product at all prior to resale, but merely add to its value by incorporating some form of service which is of benefit to the consequent purchaser, *eg* fresh prawns sold to a grocery chain which repackages the product into smaller cartons for resale in the retail outlets.

It is felt by some that the government market should be differentiated from industrial markets on the grounds that public sector activities often do not include any tangible return from their purchases. Nevertheless a public sector acquisition usually involves some form of product modification which increases the value of the original purchase. The fact that this increased value may not be recovered upon further usage of the raw material is really only an example of poor management practice.

In approaching any new market situation, the marketer needs to know the objectives of the customer because there is a key difference in the nature

27

of the buying behaviour in the two types of market. The core of the marketing concept is to comprehend the nature of demand for the company's product. This demand function and the effectiveness of the company's ability to satisfy need is closely linked to the behaviour of the customer. Consumer markets are typified by individuals acting to satisfy their personal values and expectations, whereas in the industrial market the purchase is linked to an economic objective of utilizing the product to generate a tangible return. In theory a more logical, less emotive behaviour pattern can be expected in the latter situation. Whichever market is being studied, however, the marketing function cannot be executed effectively until the behaviour of the potential customer is understood.

Consumer behaviour

Despite considerable research, no theory yet exists which can give the marketer a total understanding of the relationship between consumer purchase behaviour and the influence on the marketing mix, *ie* product, price, promotion and place. Attempts have been made by economists, sociologists, psychologists and biologists to produce a global explanation of behaviour but without success.

In the world of business, therefore, one is forced to accept the more practical approach of what is feasible to measure in terms of consumer behaviour and how this can be related to the purchase decision for a specific product. One of the reasons for the immense difficulty in the measurement process is that the buying activity involves five different roles, which may or may not be the responsibility of one individual. These are:

(a) The 'initiator' who suggests the possibility of the product.
(b) The 'influencer' who has some input on the final choice.
(c) The 'decider' who ultimately makes the purchase decision of how, what, when and where to buy.
(d) The 'purchaser' who is actually involved in the transaction.
(e) The 'user' who utilizes the product.

In an attempt to understand the processes involved, many marketers develop a generalized model of the purchase activity. One version of such a model is provided in *Fig 3·1* which assumes five components:

(a) Identification of need.
(b) Acquisition of information on potential products.
(c) Evaluation of the alternative propositions.

(*d*) The purchase decision.
(*e*) Post purchase evaluation of the product.

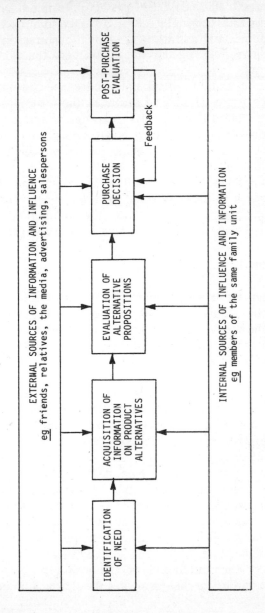

Fig 3·1   Generalized model of consumer purchase behaviour

Fortunately for the fish marketer, the basic need arousal for many products in consumer markets is hunger. This need is relatively constant and the actual mechanisms of need satisfaction using a specific product can be gained from research. The information search process can also be studied although in many instances in the fish business one can assume advertising and personal sales calls are the primary source of data for the potential customer.

Any consumer evaluating the potential purchase alternative brings to the activity a set of pre-established attitudes about the product. Having weighed these preconceptions against the range of actual alternatives, a purchase decision is made. Continued usage of the product is usual for fish products and this will cause post purchase evaluation which can positively or adversely influence the next purchase.

Given the requirement to understand the purchase model, the marketer can initiate research to measure many of the elements within it. These measurements can be used to evaluate the optimal marketing effort needed to ensure the consumer's preference for the company's product.

Market research

As mentioned earlier, the risk associated with any business decision can be reduced by the acquisition of information on the variables involved. The collection and analysis of data is achieved through market research, but the mechanism of research can be complex and expensive. It is therefore useful to adopt a simple decision rule that research should not be executed if the cost of the study exceeds the value expected from the business action which the research is designed to assist. For example, if one is planning to establish a small catfish farm capable of generating an annual net profit of £10,000 a year, there would be little point in accepting a consultant's proposal costing £50,000 to research the size and structure of the catfish market.

In developing a market research plan, it should be recognized that the cost of acquiring 'secondary information', *ie* information that has already been published, is usually significantly lower than a study to generate new or 'primary' information. Hence the first step in market research, having established the nature of the problem to be studied, is to initiate a search for secondary information. Data from sources such as financial records, sales statistics, production figures or previous reports developed for management on earlier marketing problems can be obtained from the company sponsoring the project. External sources of secondary information available to the researcher are trade association publications, government published statistics (which range across virtually every aspect of economic activity within a country), articles in business publications, the financial records of public companies and information in reviews by industry service groups such as banks, Chambers of Commerce and professional institutions.

A key objective in market research is the minimization of error. Even when secondary information is located which casts light on the research problem, it is necessary critically to appraise the nature of the data provided. If, for example, one was trying to evaluate the market attitude to the quality of scallops from various fisheries around the world, a report on the subject by a Canadian trade association might contain insufficient impartiality when commenting on Canadian production. Another possible error source can come from invalid measurements. The United Nations publishes annual export statistics for all producer countries. The method of collecting these figures does vary and hence the level of accuracy is sometimes doubtful.

When no source of existing information is located, it will be necessary to collect original data. This can be done by observation, experimentation or surveys. Of the three choices, observation can probably provide the least information on consumer purchase behaviour because the state of mind and buying motives are rarely revealed to the observer. In addition, purchasing is typically executed in the normal consumer buying environment of a retail store where one has no control over extraneous variables, eg the availability of sales people to serve all the customers on a timely basis, which can modify purchase behaviour.

The other extreme is experiment, where selected stimuli are introduced into a controlled situation and the effects of varying the stimuli are measured. The purpose of the control is to provide a relative measurement of 'normality' against which the results from the experimental situation can be measured. A retailer may only sell canned fish and wish to evaluate the potential for adding fresh fish. In the control stores no change would occur, while in the 'experimental stores' the fresh fish section would be installed. After a defined period the retailer would then compare total sales, fresh fish sales and canned fish sales in the experimental stores with those in the control stores. If there is a positive improvement in sales in the experimental stores relative to the control, the research findings indicate that addition of a fresh fish section would be beneficial to his operation.

Without any control, the retailer would not be able to decide if the increased sales in the experimental stores were due to the addition of fresh fish or some other external variable such as seasonal consumer buying behaviour.

Data collection by experiment can only provide a limited amount of information and there are significant potential errors in the design of the experiment or in the analysis which can influence the conclusions reached. Surveys can usually provide a much wider breadth of information, especially in the areas of measuring the attitudes and motives of the consumer in reaching a purchase decision.

All three research approaches require a technique to generate information. In the case of surveys, the usual approach is through the medium of a questionnaire. To generate an accurate response, the preparation of this research tool requires a high level of skill. The questions can be posed in an open ended form where the respondents answer in their own words, eg 'describe how you feel about oysters'. The answers will provide a vast range of comments, but the magnitude of the response will complicate the analysis. A closed ended question presents the answer to the respondent either in a very restricted 'dichotomous' form, eg 'Do you like oysters? Answer yes or no', or through multiple choice, eg 'Which of the following six phrases best describe your feelings about oysters?'.

To obtain the desired response a sampling plan will have to be developed. This plan will define (a) who is to be surveyed (the sampling unit), (b) how many people will be surveyed (sample size), (c) how they are to be selected (sample procedure), and (d) how they are to be contacted (sample method).

Typically one defines the sample unit as those individuals known to purchase (or influence purchase) and utilize the product. The larger the sample size, the more reliable the result. However this level of reliability has to be weighed against the higher cost of this sample versus a smaller, less accurate sample, based on fewer respondents. There is a risk of error associated with any sample which only measures response from a limited number of individuals within a population. Random sampling, ie a sample in which every member of the population has an equal probability of being selected for questioning, provides a more accurate assessment of confidence limits for the collected data. The cost of random sampling can be very high, which is why much market research is executed using a non-random (or non-probability) sample procedure. An example of the latter approach would be for a company interested in evaluating the potential for selling fish to restaurants to visit only local establishments in order to save time and money during the survey. A random sample in this instance would require respondent selection of restaurants from a list of all restaurants in a geographic area and visiting those selected no matter what the cost or time involved.

Respondents may be contacted by telephone, mail or personal interview. Although telephone interviewing is probably the fastest way to collect data, in many countries telephones are not sufficiently universal for the researcher to reach a large number of the sample units. The unwillingness of most people to spend more than a short time on the telephone also limits the potential length of the questionnaire.

A mailed questionnaire can be used to reach sample populations as long as the literacy level in the country is high enough to permit response. Again however the questions usually have to be short and simply worded. One can also encounter the problem that the rate of questionnaire return is low and/or slow. Personal interviews are the most versatile sampling method because the interviewer can probe in depth, use supplementary questions and observe reaction. Unfortunately this method is also the most expensive and requires a high level of administrative skill if more than one interviewer is involved in the programme.

Raw information from a survey will usually require analysis prior to any general conclusion being reached. If the results are in a numeric form, then statistical tests can be applied to assess the confidence limits which can be attributed to the data.

Market research is a very skilled area of marketing management which usually can be executed only by employing a market research company. In order to maximize the benefits of using an outside organization (or even if one decided to execute an intra-company study), the marketing manager is urged to read more fully a number of the available texts on market research practices. In order to illustrate the application of research, however, a hypothetical case is presented in an appendix to this chapter.

Industrial markets

The more rational behaviour in the industrial market simplifies the process of understanding why the customer reaches a specific purchase decision. However, the study of the actual buying process is complicated because in the case of a major purchase or the purchase of a frequently used product from a new supplier, a number of individuals in the customer organization are often involved.

In marketing a product under these circumstances it is, therefore, necessary to identify the responsible individuals in the customer company so that the selling claim can be made to them. To illustrate this concept, let us assume that a major fish smoker has decided to enter new markets. Their current suppliers of raw herring fillets have already indicated that they are unable to meet any additional demand for raw material. The company must, therefore, obtain fish from other sources. This new purchase activity is described (*Fig 3·2*) in a buying system based upon the generalized model introduced earlier.

The two individuals at stage one who have identified the need to locate new supplies are the Managing Director and the Production Manager.

At stage two the Managing Director and the Procurement Manager, drawing from their own knowledge of the fishing industry and from information provided by primary producers, can gather the necessary information on new sources of supply. The key issues involved in the evaluation of alternatives are the quality of the raw material, continuity of supply, and relative prices. Stage three will, therefore, require more input from primary producers as well as samples for examination. The technical issues of raw material demands will require involvement by the Production Manager and Quality Control. The Procurement Manager will comment on availability claims and price quotes. The actual purchase decision is a vital subject and will probably involve all the members of the group and quite possibly will also require confirmation of previous statements by the primary producers who are short-listed as potential suppliers.

Once the new fish has passed through the smoking process, the Production Manager and Quality Control will re-evaluate their decision which will act as a feedback on whether everybody is satisfied.

It is apparent from even this simple model that the Sales Director for a primary producer of herring will have to communicate with a number of individuals in the potential customer company if he is to succeed in obtaining an order for fish.

Given the other responsibilities of the individuals in the customer company, repeat purchases from suppliers will be handled on a much more routine basis with the Procurement Manager being the sole individual involved in the purchase decision. Nevertheless he will be

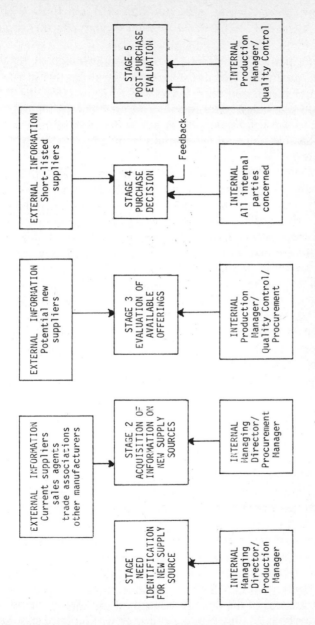

Fig 3.2   Purchase behaviour model for a fish smoker

35

influenced to some extent by post-purchase evaluation mechanisms such as quality control reports. Hence it is vital for the producer Sales Director to retain communication links with members of this department in the customer company. Even if one is not an approved supplier and the customer is apparently not considering new purchases, it can often be from informal discussion with Quality Control and/or the Production Manager that one hears of a raw material problem which your company may be able to solve. By raising this issue with the Procurement Manager, it may be feasible to be added to the list of approved suppliers for the repeat business.

Constructing a buying model for a particular industrial situation is relatively simple and does not require more than minimal research expertise to acquire the necessary data. Unstructured interviews with individuals from other companies already supplying the potential customer can be arranged, or in many cases an informal meeting with the customer Procurement Manager will reveal when and how other members of the company participate in the buying process.

Although it has been stated that industrial behaviour is rational, it is well to remember that in the selling process we are still dealing with individuals. Where there is little to differentiate between suppliers in terms of quality, price or service, the industrial buyer has less basis for rational choice. Since he has various alternatives by which to satisfy the company procurement objectives, he can be biased by personal motives. Under these circumstances the order is more likely to be placed with the salesman the industrial buyer knows and trusts than with a company with which he has had little prior contact.

Appendix

The salmon canners association case

A salmon canners association has become increasingly concerned about declining sales of salmon relative to other, lower cost, species of canned fish. It was decided at the annual convention that the association's advertising agency should be commissioned to develop a campaign to stimulate awareness of the benefits of canned salmon.

The agency immediately pointed out that without any information on consumer purchase behaviour it would be difficult to develop an effective campaign. Therefore, market research would be needed to obtain information on consumer attitudes and usage patterns for the product.

It was decided that the sampling unit should be housewives because they are involved in virtually every aspect of the purchase decision from identification of need through to post-purchase evaluation. Previous experience had indicated that with this type of research, a sample of 2·5% of the sample unit population will provide an acceptable degree of accuracy. In order to minimize non-response error, *ie* individuals not willing to respond to questions, the list of respondents was randomly selected from a larger list of households known to be willing to participate in research studies. (It should be recognised that a sample error problem may be encountered in that the people who are unwilling to respond may also have a different set of opinions, which of course one will be unable to measure.)

As consumers would be asked to complete a detailed questionnaire to evaluate attitudes and their food usage patterns, it was felt the data could most effectively be collected using a mail survey. Because there was some belief that consumption of salmon varied by region, the number of mailings to each of the four regions of the country was made in direct proportion to the population living in each region.

As little information existed on consumer attitude to fish products, it would be necessary to gain some understanding of the relevant issues before a suitable questionnaire could be developed on salmon. Professional psychologists were hired to initiate a series of interviews with consumers to probe such matters as why do people eat fish, what types of fish do people eat, what real and symbolic problems do they associate with fish. Forty people (ten women, ten men, ten teenagers, ten children) were located to participate in individual depth interviews. The discussions were relatively unstructured with the interviewer free to proceed into any area which appeared relevant. This research provided sufficient grounds for generalizations about consumer attitudes to fish.

Sixteen group interviews, four groups per city across the country, were arranged with each group comprising eight women (two heavy users of fish age group 20–40, two heavy users age group over 40, two light users age group 20–40 and two light users age group over 40). The trained moderator at each meeting posed questions based upon concepts evolved from the previous individual interviews.

This explanatory research was then used to generate certain hypotheses

which would be verified quantitively by a consumer mail questionnaire sent to 1,500 housewives using the sampling plan described above. The survey attempted to measure the following issues:
— usage pattern of salmon and other fish species
— experience of various species of fish
— attitude towards canned salmon
— the type of housewives who buy salmon (who are they, where do they live, their family characteristics, income and age)
— how often is canned salmon purchased
— what other types of fish are purchased
— how is canned salmon prepared
— who eats canned salmon, when and with what

The returned questionnaires provided data on the usage pattern for canned salmon (see *Table 3·1* for selected results) which indicated a higher proportion of triers (96%) than for the other species of fish. Trial was greater amongst higher income groups but showed little variation in relation to age or family size. The number of triers was greater in the two regions near the coast, and lower in the other two inland areas. Less than half the population served the item on a regular weekly basis, although over 71% used it at least once a month.

Information on attitude (see *Table 3·2* for selected results) indicated major variance between regular users and other groups in the area of the variety of ways to prepare canned salmon, usage of leftovers and ease of preparation. In the other attitude areas, there were only minor differences of opinion except over the issue of 'inexpensive'. On the basis of this research, the advertising agency concluded that many people perceived salmon as not being able to provide the raw materials for a variety of different meals. Furthermore, they believed the irregular users might change their opinion on the 'inexpensive' issue if they realized how flexible the product is in terms of a variety of menu ideas and in using up leftovers. They therefore recommended an advertising campaign that emphasized all the different ways in which salmon can be prepared as a family meal, to be linked with an in-store merchandising campaign by major super-markets based upon the same theme.

Table 3·1 SELECTED DATA ON USAGE OF CANNED SALMON*

Item : Trial	% Response				
	National	Reg 1	Reg 2	Reg 3	Reg 4
TRIAL RATE SALMON					
Tried last 12 months	84	89	81	83	94
Ever tried	96	98	94	95	99
Never tried	4	2	6	5	1
TRIAL RATE OTHER SPECIES					
Tried last 12 months	64	71	58	62	75
Ever tried	76	79	69	74	83
Never tried	23	21	31	26	17

Item : Characteristics of Last 12 Months Triers

	National	Reg 1	Reg 2	Reg 3	Reg 4
FAMILY INCOME					
Less than 4,000	65	69	61	64	63
4,000 – 10,000	86	98	82	84	86
Over 10,000	90	92	87	88	89
AGE OF HOUSEWIFE					
Under 35	86	87	80	83	89
35 – 50	86	86	71	87	87
50 and over	84	84	83	70	88
FAMILY SIZE					
No children	85	88	82	84	88
1 – 3 children	85	84	83	85	89
More than 3 children	83	82	84	82	87
FREQUENCY OF SERVING					
At least once a week	39	46	37	34	49
At least once a month	71	81	62	61	73
Less than once a month	74	83	67	63	76

*Hypothetical data

39

Table 3·2 SELECTED DATA ON CONSUMER ATTITUDE
TO CANNED SALMON*

Statement	% Response		
	Serve at least once a week	*Serve at least once a month*	*Less than once a month*
Know many ways to prepare canned salmon	82	68	46
Leftovers can be easily used	78	57	39
Almost everybody likes salmon	67	64	54
Provides lots of protein	45	46	42
Inexpensive	52	40	23
Always tastes good	36	34	30
Low in cholesterol	28	24	25
Good for dieters	26	21	19
Easy to digest	31	30	30

*Hypothetical data

4
Market segmentation and strategy

Segmentation

Having adopted the marketing orientated philosophy, the traditional corporate approach has been to assume that the market is composed of homogeneous needs and to introduce a single product which appeals to the largest number of buyers. This is often referred to as 'mass' or 'undifferentiated' marketing, and is still widely used by the fishing industry. A number of companies in other industries, *eg* General Motors, Coca Cola, intuitively recognized the opportunity for increasing sales by the introduction of a range of products, thereby meeting more effectively the diversity of buyer demand. The obvious advantage of having a more diversified product line has also caused a number of fish companies to move into a broad range of different species and product forms, *eg* breaded fish, batter coated shellfish, seafood entrées.

Product diversification or 'differentiation' will usually generate increased sales but it will be accompanied by higher operating costs due to one or more of the following:

(*a*) Production costs: In most situations a factory producing one item will be more cost effective than a factory which is manufacturing a number of different items, for in the latter instance production capacity may be lost during the periods when the processing line is being changed from one item to another.

(*b*) Inventory costs: To avoid 'out-of-stocks' a company will often carry safety stock to cover unexpected variations in demand. The more items marketed, the greater will be the investment in the safety stocks for all items in the product line.

(*c*) Promotion costs: Differentiated marketing can involve a range of marketing programmes to support various products sold by the

41

company. In this situation, the sum of the individual programmes will exceed that of a company which is a 'mass marketer' of a single product.

The major financial implications of differentiated marketing have led to the concept that a more cost effective management approach could be, first, to attempt to divide any market into homogeneous categories of customers (or segments) and to select specific target segments that can be exploited using a uniquely appropriate marketing mix. This technique, known as marketing segmentation, is especially appealing to a company with limited resources. For instead of going after a small share of a large market, one specific market segment can be chosen and the domination of this area of buyer need can result in the virtual exclusion of any competitor. By concentrated marketing, the company can specialize to the point of achieving operating economies in such areas as production, promotion or distribution and thereby maximize the rate of return on company assets.

To illustrate the concept of market segmentation, let it be assumed that research on buyer needs for farm raised trout isolated two distinct characteristics — price and 'quality/portion control', ie high quality products available in carefully graded packs containing minimal variation in the specified weight of each fish. A plot of buyer opinion for these two parameters could reveal one of three possible outcomes (*Fig 4·1*). In two of the situations, the homogeneous and the diffuse preference markets, there is no opportunity for segmentation. However, in the case of the clustered preference market, a trout farm could decide to modify the company's rearing and processing procedures to specialize in one or more of the three market segments. If competitors then continued to offer one general product form to all buyers, their output would soon be perceived as unable to satisfy the three distinctly different market needs.

To execute effectively a market segmentation strategy, it is necessary that one can measure the characteristics of specific buyers. The defined segments must then be accessible to exploitation through the company marketing mix. For example, manufacturers of breaded sole fillets may state a preference for an exact weight of raw fillet. Primary producers could in theory supply such an item, but the cost of selecting one exact size from the range of fillets available during processing would in most situations result in a higher selling price than the breaded fish manufacturers would be willing to pay. In addition, the identified segment

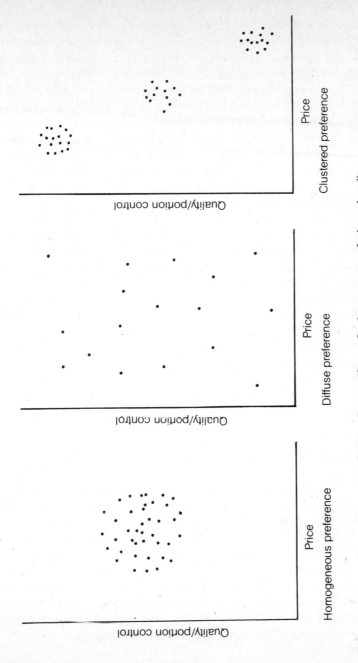

Fig 4·1 Buyer preference for the parameters of price and quality

43

must be of a size sufficient to justify investment in the development and execution of a specific marketing programme to exploit that segment.

There are numerous variables which can provide the basis for dividing a market into specific subsections when a company is considering the possibilities for market segmentation. The four most widely used are geography, demography, psychography and buyer behaviour.

The earliest form of segmentation was geographic, where a company identified regions of the country in which they could exploit a competitive advantage. An aquaculture group specializing in fresh fish may decide to locate production units in areas that, within a 50-mile radius of each unit, contain at least five population concentrations with *per capita* income levels sufficient to support a healthy restaurant industry.

Demographic variables such as income, age, sex, family size, social class, and occupation are relatively easy to measure. Hence a popular mechanism of segmentation is to attempt to correlate such variables with specific product usage. Once such relationships have been established, one can use specialized media, *eg* a certain type of newspaper or magazine, to communicate the company's advertising message. A company which markets live lobsters using a home delivery programme may decide that the majority of their customers are professional couples in the age group 35–55 with no children. The company can then select magazines with a suitable readership profile in which to advertise their product.

The complexities and unpredictability of human behaviour have forced marketers to conclude that certain products cannot be segmented using demographics. An alternative has been the use of psychographics such as 'life style' factors to identify the segment of interest. The lobster company, for example, may find that a better definition of potential customers is those who take expensive vacations, hold frequent dinner parties and play sports on a regular basis.

Another behavioural approach for segmentation is related to identifying product usage characteristics in the population. It may be possible to divide the market into non-users, potential users and regular users. The regular users will probably react to a very different marketing mix than that needed to attract potential or non-users. A very similar approach is to divide the market by usage rate into heavy, medium and light users. The hope is that each user type has specific personal characteristics and media habits which would permit the use of selective advertising campaigns to appeal to variable usage behaviour patterns. The obvious objective in this situation is to convert all buyers to a heavy usage pattern.

Strategy formulation

Once a company has decided on a business objective, typically expressed in the form of financial targets for sales and net profits, it is necessary to develop a marketing strategy to define how the company can achieve these objectives.

Step one is to decide whether the most appropriate mechanism is un-differentiated, differentiated or a concentrated marketing approach. Whichever is chosen, it should be recognized that segmentation of the market is usually achievable and that a company which offers specific products to meet specific needs will have a greater probability of success.

For a segment to be of appeal it must be of sufficient size, preferably have potential for further growth, not be over-occupied by competition and have an identified need which the company is uniquely capable of satisfying. A company which owns five inshore boats landing a range of groundfish species on virtually a year-round basis would have a number of potential outlets through which to market the catch. The manager may identify that local restaurants cannot obtain adequate supplies of fresh, consistent quality fillets and hence decide to concentrate on this market segment in the near future.

If a company has no experience of the chosen segment, then it will be necessary to decide upon a market entry strategy. This can be achieved through internal development using the existing skills of the management team, by collaborating with another company, or by acquiring an existing company already operating within the chosen segment.

A further element in the marketing strategy is for the company to determine how the variables of the marketing mix (product, price, promotion and distribution) can be utilized to influence buyer behaviour. One of the fundamental product decisions is the quality of the item to be offered. Price is one of the key information sources utilized by buyers to evaluate quality. In order to formulate the strategy on these two variables, the relationship between them can be examined through a quality/price matrix (*Fig 4.2*).

The greatest risk propositions lie in the choice in which the product quality may be seen by the potential buyer as incompatible with the price. One can expect extreme suspicion for any of the mixed attribute concepts, *eg* high quality/low price, and the more likely market acceptance will be for those co-ordinations along the matrix diagonal from top left to bottom right.

Promotion is the variable through which information on the product

		PRICE		
		High	Average	Low
P	High	Premium price strategy	Market penetration strategy	Value for money strategy
R				
O				
D				
U				
C				
T	Average	Market skimming strategy	Average market position strategy	Economy strategy
Q				
U				
A				
L				
I	Low	Single sale strategy	Inferior goods strategy	Cheap goods strategy
T				
Y				

Fig 4·2a Quality/price matrix

claims are communicated to the market. If a small number of buyers dominate the market, *eg* the procurement managers for frozen food manufacturers who buy raw materials from primary producers, these can be reached using a sales force. Where the market is comprised of individuals all making small unit purchases, *eg* housewives buying fish for their families, then advertising will probably be the most cost effective promotional device.

Managing the flow of goods from the producing company to the final market (or distribution management) is related to such factors as the seasonal availability of the products, the geographic distance between plant and point of purchase, and the buying behaviour of the final purchaser.

In the fishing industry, where the majority of primary producers are located many miles from the final market, distribution can be very important. The fact that fish is an extremely perishable commodity further complicates the distribution process.

The concept of marketing strategy to define the purpose and direction

		PRICE (£/lb)	
	High	Average	Low
Premium grade raw material	5·00	4·50	4·00
Inspected for zero sand veins			
Tail count ±5 from specification			
Layer packed			
Careful glazing			
Medium grade raw material	4·50	4·00	3·50
Occasional sand veins			
Tail count ±10 from specification			
Boxed in wax container			
Reasonable glaze level			
Low grade material	4·00	3·50	3·00
Excessive sand veins			
Tail count ±20 from specification			
Packed loose in bags			
Excessive glaze			

(Left margin vertical labels: P R O D U C T   Q U A L I T Y)

Fig 4·2b Quality/price matrix for frozen peeled and de-veined prawns

of corporate plans cannot be overstressed. Unfortunately, the sub-components of the strategy (market segment choice, market entry mechanism and marketing mix alternatives) tend to be interactive variables which can complicate the planning process. Illustration of this fact is provided by examining the earlier example of the small processor who decided to market his catch as fresh fillets to local restaurants. Up to this point the boats have been selling their fish to a local wholesaler who has accepted all responsibility for marketing. Hence our example company has no experience of selling to restaurants and needs to reach a decision on how best to enter this new market. Even if they prefer the concept of internal development of the necessary marketing skills, they will still face the issue of how to deliver the product. Certainly the local fish wholesaler will not be prepared to assist in the creation of what he will perceive as a new source of competition. Even if our company can afford the investment in a fleet of trucks to deliver to restaurants, it would probably be uneconomic to have these vehicles only delivering the typical daily needs of 20–40lbs of fish to each customer outlet.

After considering this issue, the company decided that a joint venture with another group already experienced in marketing perishable goods would be the best market entry strategy. They therefore opened negotiations with a local meat processing plant which has a van salesman/delivery system already in operation. The agreed programme would be that the fish company would hire a sales manager and he would work with the meat company training their staff on how to market fish products. In return for this role, the meat company would receive a commission on all fish sales in the area.

Further study of the fish market revealed that the restaurant owners emphasize premium quality as the key component in their purchase decision. The company is very capable of supplying such fish and decided that a premium quality offering can be most convincingly presented if marketed at a premium price.

The marketing mix issue is relatively simple because the concentrated buying behaviour in the market will cause personal selling to be the main mechanism for communicating the product story. The distribution issue has already been evaluated because it was a component of the early deliberations over market entry.

No strategy is of benefit to a company unless an attempt is made after implementation to measure the results of the programme through the use of a control system. Thus the original plan should contain a quantified statement of expected sales per customer, the number of customers, the marketing costs of the programme, and the expected net profit. Then within only a short period the organization can examine actual results and determine whether the strategy on market position, market entry and marketing mix is correct or if any revision is necessary to ensure attainment of the required corporate objectives.

5
Product policy

The most crucial component of the marketing effort is the product. It is the basis for customer need satisfaction and generates the revenue from which all of the company activities are funded.

As potential customers usually face numerous purchase alternatives, the company is rarely in a monopoly situation. It is therefore necessary that the company establish a unique identity for the product, using characteristics, or attributes, that can provide a basis for differentiation from competition in the market. An indistinct product is unlikely to survive.

In order to reach a decision on the nature of differentiation the company will need to determine:

(a) The attributes that are important to the customer.
(b) The perceived positioning of competitive products.
(c) What is the best position or 'product space' for the company to occupy

A product consists of a wide range of features which include price, texture, name, availability and quality, all of which can provide the basis for differentiation. In most circumstances, however, the potential buyer will utilize only some of these attributes to reach a purchase choice decision. The relative importance of the attributes considered by the purchaser can be measured using a variety of research techniques. One of the simplest is an attribute and usage study of the type described in the canned salmon example. The weakness of this approach is that the results cannot easily be used to identify the relationship between important attributes and the product positionings. To achieve this form of evaluation it will be necessary to use multi-decisional scaling where the respondents are asked to rank different products for each of the chosen attributes. The results can then be combined to form a map of product space which will provide a visual description of positionings.

To illustrate the mapping technique, North American procurement managers purchasing fish blocks for use in the manufacture of breaded fish portions could be asked to rank various species in relation to the attributes of cost and flesh quality. Their responses can be plotted on a graph using the two attributes as the axes (*Fig 5·1*). It is apparent that species such as cod and haddock are considered to provide high flesh quality at a cost higher than most other species. An extreme contrast to this situation is minced cod and Alaska pollack which are rated low for both quality and cost.

The same respondents can be asked to specify the attribute ranking for an ideal species, and the density of their preferences described in the form of boundary circles on the map of product space. For the market segment requiring high quality raw material, the boundary plot for the ideal species will contain haddock and cod. In the economy segment of the market, procurement managers will forego flesh quality to meet the minimum cost required for the finished product. The ideal species boundary encompasses various species of minced fish. The third ideal species which can be described as the 'reasonable quality/reasonable price' segment would appear to only just encompass the apparent position for whiting and Alaska pollack.

The product space map and the ideal species plots can be used both to examine the current position of various items and to evaluate the opportunities for new products which might be introduced. For example, farm raised tilapia would probably receive a low rating for flesh quality and hence a company marketing blocks made from this fish should only consider the economy segment as a potential positioning.

For established species, the potential can be reviewed for a change in flesh quality and/or price. Deep skin whiting is an example of a species where a change in the processing technology radically upgraded flesh quality and resulted in the more effective fulfilment of the ideal requirements in the reasonable quality/reasonable price segment. Prior to this innovation, procurement managers either had to accept (*a*) lower price/lower quality regular whiting and Alaska pollack, or (*b*) higher priced cod blocks, as the raw material from which medium price/medium quality breaded fish portions could be manufactured. Neither choice represented an ideal solution because it was a compromise on either cost or quality standards to fulfil buyer demands in this very large market segment.

50

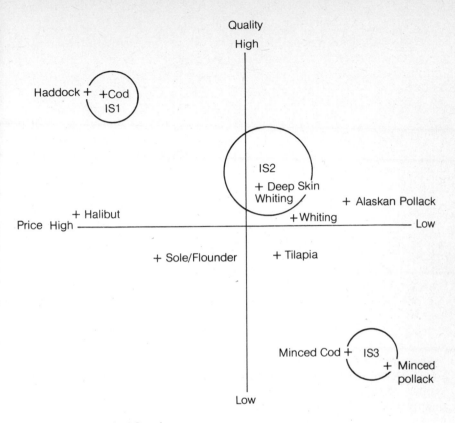

IS — Ideal Species
IS1 = High Quality Premium Species Preference Space
IS2 = Reasonable/Quality/Reasonable Price Preference Space
IS3 = Economy Species Preference Space

Fig 5·1   Product space map

Product life cycle

A product space map is never static because products exhibit a life cycle related to their degree of acceptance within the market. The life cycle is usually characterized as a three-stage S-shaped curve describing the phases of introduction, growth and maturity. This will eventually be followed by a decline phase with sales moving towards zero (*Fig 5·2*).

When a product is first introduced, market acceptance will be low

51

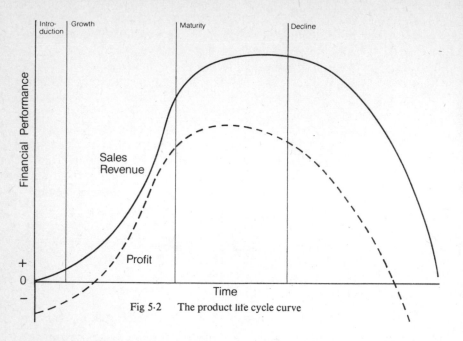

| Intro-duction | Growth | Maturity | Decline |

Financial Performance

Sales
Revenue

Profit

+
0
−

Time

Fig 5·2   The product life cycle curve

because potential customers will be purchasing other items to satisfy their needs. It will be necessary to create a change in buying behaviour using promotional programmes to generate trial and build awareness. Once the new product has achieved an initial trial level, growth in sales can be expected as repeat usage begins to develop and additional new trial occurs. Eventually the number of new users will start to fall and a steady state repeat business will evolve, bringing the product to the maturity stage in the life cycle.

Fisheries biologists may use the Lotka-Volterra predation model to define the typical S-shaped curve associated with the rate of change in a fish population over time as more vessels enter the fishery. If the fishery is well managed a steady state yield is established which reflects the carrying capacity of the environment and a recruitment rate equal to the combined effects of natural and fishing mortality.

This population model can have direct influence on the product life cycle, for when a new species is introduced to the market or a new fishery opened for an existing species, market acceptance may be low. Hence the market demand would not support more than a minimal level of landings. As market capacity grows more fish can be sold and fishing effort will

52

increase to fulfil sales as the product moves into the growth phase of the life cycle. At the point of maturity on the life cycle curve, the population dynamics of the fishery can influence events in three ways:

(a) Maximum potential landings at equilibrium in excess of market demand. Landings continue to rise beyond the market size at maturity. Supply can exceed demand; fishermen may use price reductions to sustain sales volume whereupon some fishermen could be expected to leave the fishery because production costs will exceed the lowered selling prices. This will cause some oscillation in the landings as fishermen enter and leave the fishery until an equilibrium between market demand at maturity and supply is achieved.

(b) Maximum potential landings at equilibrium less than market demand. In the event that the landings at the equilibrium point are less than market demand at maturity, sales volume will stabilize at a level lower than the potential market size for the product if the fishery is well managed. In these circumstances there will be a 'gap' between the volume of supply and the volume of demand.

(c) Overfishing to the point of irrevocable damage to the fish population. If overfishing occurs and landings begin to decline drastically, then actual sales will be significantly lower than the potential at the maturity point on the product life cycle. If overfishing is permitted to continue unabated, fishermen may eventually be forced to abandon the fishery because of excessive costs per unit of catch. Landings will trend towards zero and the product life cycle will move into the decline phase because of the unavailability of product.

The same product life cycle concept can be expected in the development of a market for a new species introduced by the aquaculture industry. Initial market acceptance will be low until widescale customer acceptance can be created. As demand increases new culture units will enter the industry and capacity expansion can occur up to the point of market maturity on the product life cycle. Beyond this level of demand, additional aquaculture output would be inadvisable because it would be in excess of potential total market sales.

Unfortunately, excessive output has already occurred on a number of occasions in the past twenty years. The United Kingdom trout farmers recently overproduced and they reacted by reducing prices to a point where the less cost-effective operators marketed below their product cost. This inevitably led to bankruptcy for some farms and a revision of the industry structure more in line with market demand.

One advantage aquaculture has over the traditional fishing industry is that culture operations will not usually face a situation where total possible output is lower than market potential. In fact when culture units are considering the choice of species, it may be advisable to include an evaluation of the supply 'gap' between landings from traditional fisheries and market sales potential. Where there is very little difference between potential sales and landings, the fish farmer would be advised to avoid the species under review because incremental production through aquaculture would rapidly force the market into excess supply, with the consequent risk of a fall in the wholesale prices.

Life cycle profitability

During the introductory phase, the ratio of marketing expenditure/sales is high because of (*a*) the need to generate customer awareness and trial which requires promotional expenditure and (*b*) the low sales base. In the growth phase, the company will continue to spend heavily on promotional activity with the objective of gaining a dominant market position before the entry of competitive products. Sales will also be rising during this phase and hence the expenditure/sales ratio will gradually fall. The effect of this situation is that profitability will peak towards the latter stages of the growth phase (*Fig 5·2*). As the market moves into maturity, the level of competition tends to increase as the supplier companies attempt to maximize their share of a static market. The participating companies engage in product improvement, market segmentation, and changes in marketing mix to generate incremental sales. Both involvement in and reaction to these activities by any one company will tend to increase the expenditure/sales ratio and lower earnings. Profitability will also begin to trend downwards when the market starts declining.

In the fishing industry, a move to increase sales volume is very typically associated with the need to invest in the expansion of catching or culture output capacity and processing facilities. It is, therefore, vital to understand the status of the product under review in relation to the current position on the life cycle curve. Sales expansion at maturity can only come from competition and the resulting marketing expense will reduce profitability. The lower profitability and capacity expansion will then jointly act to reduce return on investment which is calculated on the basis

ROI = Profit ÷ Investment.

The alternative proposition is for sales expansion of a product at the growth phase on the life cycle curve, which can be achieved through increased customer usage of the species. This will result in higher profitability and an improvement in the return on investment if incremental profits exceed the investment required to expand capacity.

The conservative nature of management in the fishing industry would seem to be the reason that many companies emphasize sales expansion when the market is already approaching maturity. During the sixties many European and North American fishing operators invested heavily in new trawlers and processing facilities in reaction to the growth in frozen fish sales. It would appear many of these investment decisions were made after the market for frozen fish had reached the maturity phase. To sustain sales of their additional output some companies then increased marketing expenditure. Because such sales could only come from competition, the cost of attracting new customers was much higher than during earlier years when the major source of sales came from attracting new users of frozen fish into the market. By the mid-seventies, this situation was reflected by a declining return on investment for primary processors which was then further compounded by rising fuel and other operating costs.

Management of decline

The decline phase of the product life cycle can occur because of (a) a change in customer buying habits, eg in a country where a rising standard of living causes consumers to switch from low cost fish to more expensive protein foods such as meat or poultry, or (b) a new technology which makes an existing product obsolete, eg the introduction of canning to preserve fish which resulted in a reduced demand for salted fish.

A major error made by many companies is that insufficient attention is given to the elimination of products which are at the decline stage in their life cycle. The management attitude is one of complacency as long as the item continues to generate sales revenue in excess of the direct costs of production. This attitude prevails because of a failure to recognize the following hidden costs associated with this situation:

(a) The products consume a disproportionate amount of management time.

(b) Production runs of a short duration involve expensive changes in line set-up.

(c) Sales volume is less predictable, requiring additional inventory coverage to avoid 'out-of-stocks'.

(d) Promotional expenditure is less effective in generating sales than if allocated to items still in a growth phase.

(e) The product can damage corporate image both among the customers and members of the distribution system.

In order to avoid this situation, the marketing group must execute a periodic performance review to identify products to be eliminated if their profitability has reached, or is approaching, a defined minimum achievement level.

In the fishing industry, onset of product decline is often caused by reduced availability of a species due to overfishing, *eg* North Sea herring stocks, or a change in environmental conditions which adversely affects stocks, *eg* debris from a volcano in Nicaragua which blanketed the benthos and led to an abrupt fall in shrimp landings. Except in the case of a catastrophic fall in catch, the onset of the decline phase can usually be predicted by examination of annual landing trends. Once an availability constraint develops, the companies affected have a number of decision paths open to them, depending upon the importance of the affected species relative to total sales. The least risk proposition is to de-emphasize the affected item with the eventual aim of elimination as the market size approaches zero. To retain or increase the importance of the product as a component of total sales is a much higher risk decision path and should only be considered if raw material availability can be guaranteed through actions such as:

(a) Gaining sole or dominant rights to the remaining stocks, *eg* Pacific halibut fishermen who have been granted licences for a fixed proportion of catch.

(b) The development of a technology to reduce the proportion of the constrained raw material in processed products, *eg* the addition of vegetable protein extenders to produce blended shellfish products.

6
New product development

The implication of the product life cycle is that eventually all products will enter the decline phase. Replacement of the lost profit contribution from falling sales can only be achieved by the introduction of new products. With the current rapid change both in technology, *eg* the increase in catch per unit effort with new trawling gear, and the business environment, *eg* the introduction of the 200-mile limit by many nations, no company can safely assume the products at maturity point of the life cycle curve will be a source of profit in the future. New product development must, therefore, be an area of the marketing function which is receiving continuous attention.

The failure rate for new items within a short time after introduction is extremely high. Consequently it is necessary that a development programme exist within the company which (*a*) maximizes the rate of idea generation and (*b*) effectively screens these ideas to ensure only the best concepts are progressed through to market launch. For the fishing industry, companies usually face the constraint that only one type of raw material is available to the operation. If a new product is to be developed, the lowest risk proposition is the use of the existing raw material and current technology to form the basis of a new concept. Other choices in a product decision matrix (*Fig 6·1*) are a higher risk proposition because they will involve new raw material and/or technology.

Idea generation for the existing raw materials should be related to a more effective satisfaction of current market needs. An example would be the move from packing bulk, ungraded fish to five pound, graded fillet packs for restaurants aiming to stabilize serving costs by portion control. If no opportunity exists for this approach, one alternative is to introduce new technology. Under these circumstances, the company would probably introduce new manufacturing equipment which uses existing

57

TECHNOLOGY

	CELL 1	CELL 2
R	Current raw material	Current raw material
A	Current technology	New technology
W	Risk level: Lower than all other cells	Risk level: Lower than C.4; Equal to C.3; Higher than C.1
M		
A		
T	CELL 3	CELL 4
E	New raw material	New raw material
R	Current technology	New technology
I	Risk level: Lower than C.4; Equal to C.2; Higher than C.1	Risk level: Higher than all other cells
A		
L		

Fig 6·1a   New product direction path decision matrix

TECHNOLOGY

	CELL 1	CELL 2
R		
A		
W	Introduce graded cod fillet packs	Introduce batter-coated cod fillets
M		
A		
T		
E	CELL 3	CELL 4
R	Introduce graded haddock and sole packs	Introduce frozen entrée packs using crab as the main ingredient
I		
A		
L		

Fig 6·1b   Example application of the direction path matrix

58

raw material as an ingredient to produce a superior performance processed product, *eg* installation of a line capable of producing seafood entrées.

In certain areas of the fishing industry it is possible to change to a new species without changing technology. This new product path is often attractive if a reduction in the supply of current raw material causes prices to rise, and to avoid a decline in market size a company can move into exploiting an alternative, lower cost species. This approach is exemplified by the Alaskan fishing industry which began also to land snow crab as market demand for king crab was affected by rapidly rising prices for the latter species. The majority of aquaculture systems, however, centre on one species and to change to another will usually require the development of new technology. The latter route is also open to traditional fishing operations, but for either group it should be recognized that the concept is a higher risk proposition than the other decision paths.

Prior to investment in product development, an initial survey on issues such as marketing, corporate resources, production compatibility and research and development requirements should be completed. A simple checklist with each factor evaluated on a numeric scale can be utilized to obtain an overall score for the new ideas as demonstrated in *Table 6·1*. If the total score is greater than a predetermined level, then the next stage in product development, namely, a preliminary sales and profit forecast, should be instigated.

The major variables influencing unit sales for a new product can be described by the simplified equation

$$S_u = (t)\,(p)\,(r)\,(u)$$

where

S_u = annual unit sales

t = the proportion of the buying universe who try the product, *ie* trial rate

p = the total buying universe

r = the proportion of the triers who adopt the product as a regularly purchased item, *ie* repeat rate

u = average annual usage rate of the regular customers

If a company has experience of similar products, data on these products can be used to develop estimates for the equation. Where no information exists, it will be necessary to research probable customer behaviour. Trial rates can be evaluated by discovering the purchase intent from a sample of the market population. Placement of the product with a panel of potential customers can then provide some understanding of the likely adoption and usage rates.

The forecasted sales curve for the new product will reflect the combined influence of trial and repeat purchases, with the former component dominating early sales and then being replaced by repeat sales as the majority component. Given an objective to reach the maximum sales volume in the shortest possible time, a key factor of interest to the marketing manager will be the rate of product trial.

Research across a large number of products has revealed the existence of a standard form of customer purchase behaviour known as the 'adoption of innovation'. This curve (*Fig 6·2*) describes the fact that different types of customers take varying periods of time before they are

Table 6·1   NEW PRODUCT EVALUATION SCHEME

	Rank score*
Market stability	
(1) Is the market expected to be a long term proposition?	....
(2) Is the available market of wide scope (either geographically or variety of users)?	....
(3) Is there any possibility of captive markets through integration?	....
(4) Is the market stable during market recession?	....
(5) Is future market growth expected to be high?	....
Product volume trend	
(1) Is the product capable of fulfilling unique demand/need of the major proportion?	....
(2) Is demand expected to exceed supply in the foreseeable future?	....
(3) Is the product immune from expected technological change?	....
(4) Does the product offer export potential?	....
Marketing factors	
(1) Will existing customers purchase the product?	....
(2) If new markets are to be entered, is entry a simple low-cost proposition?	....
(3) Will the product represent incremental growth relative to the current production line?	....
(4) Can potential medium/heavy users be identified?	....
(5) Will demand be relatively non-seasonal?	....
Sub-total carried forward	....

Sub-total brought forward   . . . .

Feasibility

(*1*) Will the product be introduced and established before competitive
reaction can occur?   . . . .

(*2*) Will the product be an exclusive corporate proposition?   . . . .

(*3*) Does the product add support to the company's overall market image?   . . . .

(*4*) Are the raw materials easily procured?   . . . .

(*5*) Does the procurement of the materials assist the negotiating position
with current suppliers?   . . . .

Research and development (R & D)

(*1*) Will development utilize current R & D skills?   . . . .

(*2*) Will acquired knowledge assist future R & D programmes?   . . . .

(*3*) Will product development assist speed of progress on other
active projects?   . . . .

Production

(*1*) Will the product consume currently unused manufacturing capacity?   . . . .

(*2*) Will the product utilize current production worker skills?   . . . .

(*3*) Will the product consume by-products from current processing
operations?   . . . .

Total score

———
* Rank score:
  Highest possible score for each factor 10, lowest 1.

  Total score to exceed for example 200, for the product to be moved to the next phase of
  development.

willing to try a new product. The process is usually represented as a
normal curve and for convenience the customers are divided into five
different types. The important characteristic of the curve is that the time
span from 'innovators' to 'laggards' is specifically related to the speed with
which the company marketing programme can generate awareness and
interest in a new product within the buying universe.

The two variables of the marketing mix through which the marketing
manager can influence awareness and interest are price and promotion.
Permutation of these variables creates four distinct new product
introduction strategies (*Fig 6·3*) of which the penetration strategy of high
promotional activity/low price will usually generate the fastest trial rate.

Fig 6·2 Diffusion of innovation curve

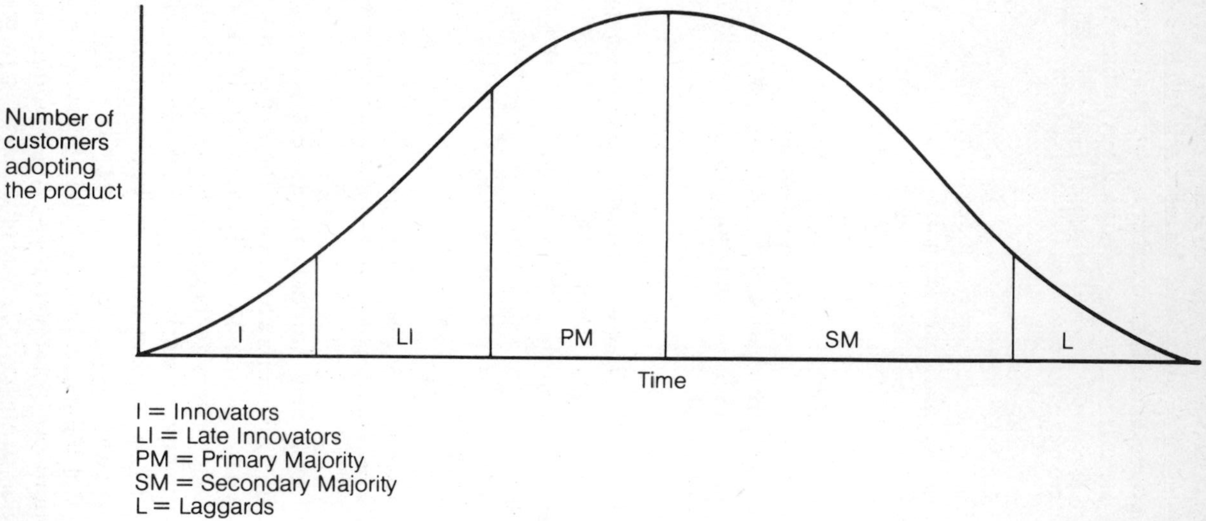
Number of customers adopting the product
I
LI
PM
SM
L
Time
I = Innovators
LI = Late Innovators
PM = Primary Majority
SM = Secondary Majority
L = Laggards

The promotion concentration strategy can also generate a high level of product awareness, but unless the market is relatively price insensitive the premium price will limit the number of potential customers. The other two strategies use minimal promotional activity, which will probably result in a slower trial rate for the new product. Hence it will take longer to reach the point of maximum sales.

Once the price/promotion strategy has been decided, the earlier sales forecast equation can be used as the basis for evaluating the financial viability for the new product, because

$$\text{annual profits} = S_u P_r - (P_t + S_u C_u)$$

$$\text{where } S_u = \text{annual unit sales}$$
$$P_r = \text{unit price}$$
$$P_t = \text{annual promotional expenditure}$$
$$C_u = \text{unit product production costs.}$$

		PROMOTIONAL ACTIVITY	
		Low	High
P R I C E	Low	Low profile economy product strategy	Market penetration strategy
L E V E L	High	Low profile market skimming strategy	Promotion concentuation strategy

Fig 6·3 Market entry decision matrix

Although some market research may have been carried out during the preparation of the sales forecast, most new products at this stage in their development process are a business idea and far from being a viable commercial proposition. The next step is to evolve from the idea to a total product concept which will survive in the market. If the product is under-going form modification after landing from a natural fishery or harvest from an aquaculture system, the production department will be involved in designing and producing a feasible product within the cost and attribute characteristics defined by the marketing group. It is not unusual to encounter problems in formulation of physical specification which will require redesign of the product concept. These changes will then have to be evaluated by preference testing amongst potential customers to ensure that the product still offers the most effective combination of characteristics versus competitive items.

Two other elements in the evolution from product idea to finished concept are naming and packaging. The choice of name will often require market research to discover the image created in the customers' minds and to determine the name most pronounceable and memorable. The chosen name should ideally communicate something about the product benefit, *eg* Cold Springs Trout, but this is not easy to achieve.

Packaging was regarded until recently as having the single function of protecting the product, at the lowest possible cost, during distribution. Many companies now realize that the packaging is a component of the total concept and in certain instances represents an integral element during final usage, *eg* 'boil-in-the-bag' fillets. Furthermore, as in most grocery stores self-service by the customer is the accepted practice, the package must be able to fulfil the promotional function previously executed by a salesperson. Therefore the package must gain attention, describe the product features, provide consumer confidence and sustain the product image already established in the consumers' minds through advertising.

As a product approaches the final concept stage, expenditure and management commitment both increase at an exponential rate. The concurrent hazard is that the marketing group focuses on any information which indicates success and discounts data supporting weaknesses in the product concept. It is mandatory that the development team have a progress appraisal system to ensure that any 'danger signals' of potential failure are recognized and acted upon immediately. During the last decade, public sector organizations in various areas of the world have focused attention on the expansion of fishing activity to include underexploited species. Once such organizations have publicly announced their opinion on the potential for a new species, it would seem that individuals involved in the product development phase are unwilling to face the fact that the original idea might not be commercially viable. Funds are expended on the assumption of success and vessels converted to exploit the resource. It is only the unwillingness of the market to accept the product that forces the final recognition of failure. The blue whiting project in the UK is a good example. Although early studies indicated potential market acceptance problems, development continued and the outcome was that some commercial companies involved in the programme received minimal or negative returns on their investment. Had progress appraisal been an accepted practice, then possibly some of the mistakes which occurred could have been avoided.

Similar, and sometimes even more costly, mistakes have been made in the aquaculture industry, with public and private sector groups continuing development on a new species or a new culture technology for an existing species despite indications of major potential problems. The high culture costs associated with carnivorous fin fish have caused a number of groups to examine exotic herbivores as alternative candidate species. Although the flesh texture and skeletal structure of these species reduces the probability of their customer acceptance, research is continuing — often accompanied by the logic that with the decline in protein supplies *per capita* of world population, a market for these fish will be guaranteed. Another example of this type of error was the continuation of culture technology research on some flatfish species in Europe. By the mid-seventies it was apparent that culture costs would greatly exceed market price, but nevertheless the work continued in the hope of a lower cost technology or achievement of a new price structure for farm-raised output. None of these solutions could have withstood a non-emotional commercially realistic examination, yet further funds were expended before participants were eventually forced to recognize the need to terminate their development programme.

One of the problems facing any management group is that even after product concept development has been completed, the ability of the new product to survive in the marketplace has still to be proved. In industrial markets, rational buyer behaviour means that if the new item has substantial advantages over competition, then in all likelihood customer trial and adoption will occur. Judgement decisions on product introductions can therefore often be made for new items in the industrial sector without further research. Within consumer markets, however, the interaction of the marketing mix variables on a less predictable customer behaviour pattern causes judgement decisions to be a high risk proposition for management. This situation has led many companies to improve their knowledge on potential sales, and evaluate alternative introduction plans, through the use of a test market.

The logic behind test marketing is to introduce the product into a limited number of small markets which have a population structure similar to that of the national market. The cost of marketing plan execution in test markets is significantly lower than for a full scale product launch. Hence the company can evaluate potential market performance without risking the level of expenditure necessary to introduce the product to the total consumer population in a country.

The key factors to be evaluated through a test market are:

(a) the actual product trial rate
(b) the level and frequency of repeat purchase
(c) the relative effectiveness of various marketing plans, eg a high versus a low weight promotional programme
(d) consumer acceptance of product benefit claims
(e) reaction of the trade to the new product
(f) potential problems of establishing an effective distribution level

This information can be collected by examining company shipment data, audits of sales rates in selected retail stores, surveys of consumer attitudes and interviews with members of the distribution channels.

A critical issue is the duration period of the test market. Certainly it should be continued until an evaluation of trial and repeat purchase can be made. If first time purchase is slow or repeat purchase frequency is low, then the test may have to continue for many months. Unfortunately, although the longer the test the greater the accuracy of the result, the company takes the risk that an extensive test schedule can permit competitors to introduce their own new items and pre-empt control of the market.

The optimistic outcome for the test market is that trial and repeat sales rates are both high. In this situation, a move to market introduction should occur immediately. If, however, trial rate is high but repeat sales are poor then it can usually be concluded that the product is not meeting customer expectations. Further product development work will be necessary. The opposite situation of low trial but high repeat sales, in most circumstances, indicates a failure to generate sufficient consumer interest in the new product. Typically this problem can be overcome by increasing the level of promotional activity. The least favourable outcome is both a low trial and repeat sales rate. This is clear indication that the item has failed to compete against existing items, and plans for market introduction should be terminated. In many instances the development team has spent years bringing the new product to this point and therefore project termination is not an easy decision to reach. For this reason, many companies have an impartial review committee to whom the project team have to justify all plans for market introduction prior to execution.

It should be recognized that the new production development process represents a major investment for the company. In order to maximize return, every effort should be made to generate incremental revenue from the product concept. The possible directions for achievement of this aim can be described by a concept expansion matrix (*Fig 6·4a* and *b*). One opportunity path is to develop new markets for the product which may be within the same country, *eg* extending a consumer product into the industrial sector, or by launching into a similar market overseas. If the marketing group lacks experience of operating in other markets, then it may be preferable to remain in the same market but develop a range of products related to the basic new product proposition, *eg* a company which firstly introduced a new smoked fillet expanding the smoking process to other species of fish or seafood. Long term the greatest potential will probably lie in a corporate expansion along the direction of both product concept and available markets.

Reliance on a single product in a single market in the fishing industry is a high risk proposition. The promotional costs of supporting a single product are usually higher per unit of sale than for a multiple product range. In fact the food industry in general is typified by low profit margins, and for many companies the only effective mechanism for funding major promotional activity is to expand the revenue base by product line diversification. The only time it is right to establish a presence in one market by this means is if overall economic conditions can be expected to sustain an on-going stable expenditure pattern by potential customers. The last decade has provided sufficient evidence that economic conditions can fluctuate violently, and one possible route to avoid this effect is to establish a corporate presence across a range of markets. The Icelandic fish industry exemplifies the corporate expansion route. In their major markets they have maximized their revenue base through offering a broad range of products. Furthermore, by operating on a multi-market basis, they have been able to switch marketing emphasis depending on market conditions; for example, in the late seventies when sales were depressed by low cod prices and adverse currency exchange conditions, effort was switched to the EEC where demand for cod remained high and currency rates more favourable.

PRODUCT

	Present Product	Improved Product	Expanded Concept	New Technology
M A R K E T Present Market	Market penetration strategy	Introduce improved performance product	Add additional products to broaden product line and increase variety of choice for customers	Examine opportunity for product performance improvement or cost reduction through new technology
New Market	Market development strategy (geographic or user type)	Extend into markets using formulation specific to user needs in each market	Market segmentation approach	Alternative technologies to suit need variation across market types
Integration	Move vertically to gain control over raw material resources and/or dominate distribution system			

Fig 6-4a   Concept expansion matrix

PRODUCT STATUS

	Present Product	Improved Product	Expanded Concept	New Technology
MARKET — Present Market	Support product with promotional programme to move to maturity phase in minimum time	Introduce boneless skinless fillet	Add other fillets to product line	Commence shipping product packed under inert gas to increase shelf life
MARKET — New Market	Enter food service market	Introduce portion control packs	Produce steaks for restaurants, lower cost range for canteen feeding operations	Produce range of breaded fish for food service wholesalers in frozen food segment

Vertical Integration — Downwards: Acquire ownership of inshore boats

Vertical Integration — Upwards: Acquire ownership of retail outlets or wholesalers

Fig 6-4b Example of concept expansion matrix for a primary processor introducing a new species into the retail market

7
Price determination

Economists have developed pricing theories to explain the achievement of equilibrium between potential buyers and sellers within the market place. Relationships between quantity and price exist on both the demand and supply sides: as unit price rises producers are usually willing to supply more produce, and conversely, as price falls, consumers are willing to purchase more produce.

This effect is an everyday occurrence in the fish industry in the establishment of achieved landing prices. When prices are high for a species, more skippers are interested in expanding catching effort and increasing supply. Buyers at auction, however, will limit purchases for a species when prices for it are high and only increase their commitments as prices trend downwards (*Fig 7·1*). The intersection point of these two variables, respectively known as the supply and demand curves, will create an equilibrium point at which supplier and buyer are in agreement on quantity and price.

This concept is not only useful for explaining the attainment of a price equilibrium, but also can be applied to assessing the possible outcome of changing circumstances within a fishery. For example, the introduction of an annual quota, as has occurred in certain herring industries, will mean that supply is fixed no matter how intense the implicit strength of market demand. The market price for the landed quota (assuming no change in the demand curve) will reflect the size of the quota. If the quota is reduced from one year to the next, for example, landed prices will be expected to rise because the supply curve has shifted to the left (*Fig 7·2*). Conversely, an increase in the landings quota would increase supply and landed prices would fall.

Alternatively, following a decision by a government body to spend funds on the creation of greater consumer awareness for a certain species,

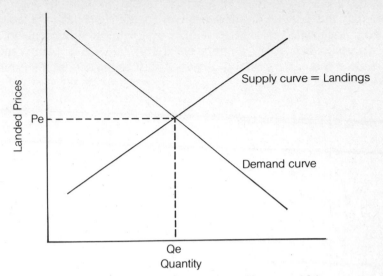

Fig 7·1   Equilibrium between supply and demand in a normal fishery

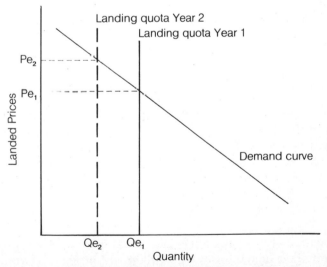

Fig 7·2   Equilibrium under fishing quota conditions

71

the demand curve will tend to move to the right with the consequent increase in market price (*Fig 7·3*). The opposite reaction occurs, of course, if a species becomes less popular due to adverse publicity such as a pollution scare: the demand curve will shift to the left and the equilibrium price will fall (*Fig 7·4*).

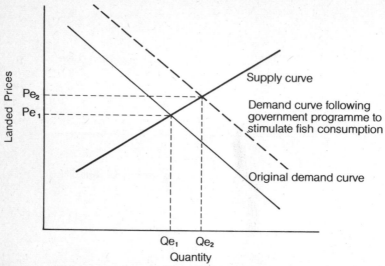

Fig 7·3   Influence on equilibrium price of programme to stimulate fish consumption

Economic models

Other work by economists has led to the development of a simple model by which a single producer can determine the optimum price for output, if the objective of the company is to maximize profits. An illustration would be the application of the model for farm-raised freshwater shrimp (*Macrobrachium*). Assume the company has carried out research and established a demand curve for their output which can be expressed by

$$Q = 500,000 - 100,000P$$

where Q = quantity sold in lbs ... ... (1)
       P = unit price in £ sterling

(The implication of this equation is that as the price approaches zero, the maximum market demand is for 500,000lbs of product. Further, as the price rises towards £5·00/lb, the quantity sold will approach zero.)

72

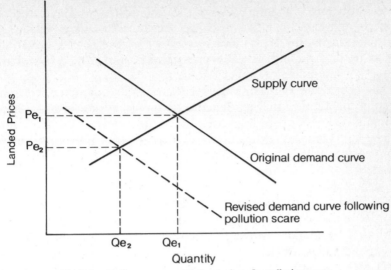

Fig 7·4   Influence on equilibrium price of a pollution scare

Sales revenue, S, can be described by the equation

S   = PQ

and net profit, NP, as

NP = PQ – C

where C   = total production costs,

and as total farm costs are comprised of a fixed cost of £100,000 plus variable costs directly related to the level of output, Q, then

$$C = 100{,}000 + Q \quad \ldots \quad \ldots \quad \ldots (2)$$

Hence NP = PQ –(100,000 + Q)

and substituting for Q in (2) using equation (1) yields:

$$
\begin{aligned}
NP &= P(500{,}000 - 100{,}000P) - (100{,}000 + 500{,}000 - 100{,}000P)\\
&= 500{,}000P - 100{,}000P^2 - 100{,}000 - 500{,}000 + 100{,}000P\\
&= 600{,}000P - 100{,}000P^2 - 600{,}000
\end{aligned}
$$

and differentiation yields

$$dN/dP = 600{,}000 - 200{,}000P$$

which gives a maximum point for profitability where

$$200{,}000P = 600{,}000$$

or where P = £3·00

at which point, Q = 200,000lbs and

$$
\begin{aligned}
\text{Net Profit} &= (3)\,(200{,}000) - (100{,}000 + 200{,}000)\\
&= 600{,}000 - 100{,}000 - 200{,}000\\
&= £300{,}000
\end{aligned}
$$

73

The economists' theoretical models assume that price is the only functional component in the purchase decision. This may be valid in a subsistence economy, but as individual wealth increases, other variables such as product quality, distinctiveness and positioning begin to have a more important influence on behaviour.

Another criticism of the models is that in practice companies are unable to obtain the data on market demand or supply curve, let alone have a precise definition of operating cost curve across various levels of output. A more important issue, however, is whether profit maximization over either the near or long-term is an actual corporate objective. It may be an ultimate goal, but in many cases more specific, immediate objectives prevail such as:

(a) Market penetration: where the company sets prices relatively low in order to maximize product trial and discourage competitive activity.

(b) Market skimming: where the company sees an opportunity to demand a higher price which may limit total sales but provide a high return per unit of sale. Companies frequently adopt this approach when they believe the immediate market volume is limited and that a price reduction can be made at a later stage if economies of scale in production result in lower unit manufacturing costs.

(c) Product line pricing: where the company may set a price giving low or minimal profits on one item to stimulate sales of other company products. This is often referred to as 'loss leader' marketing.

Therefore, although the economists' ideas on pricing decisions are useful in evaluating theoretical principles, typically in actual business price setting has to be examined in circumstances of imperfect information.

Pricing decision with imperfect information

Cost orientated pricing
Many companies calculate the price to be quoted to customers by adding an arbitrary profit target to their known production costs. For example, if the manufacturer of breaded fish fillets has calculated that the production costs for one shift of product with a 50% flesh/50% breading ratio

$$= 15,000x + 15,000y + \text{allocated overhead}$$

where

x = price/lb for fish fillets
y = price/lb of the batter and breading coating

overhead = £6,000 per 30,000lbs of production
and if

$$x = £1.20$$
$$y = £0.40$$

then
production costs

$$= £18,000 + £6,000 + £6,000$$
$$= £30,000$$

and
cost per lb

$$= £30,000 \div £30,000$$
$$= £1.00$$

If the company has a predetermined profit of £0.20/lb, then the quoted price per lb,

$$= £1.00 + £0.20$$
$$= £1.20.$$

This method is simple to apply, which is the reason for widespread acceptance across most industries. It can be criticized, however, on the grounds that it ignores the influence of the demand element of the market environment. Alternative price mark-up could alter the total sales volume for the company because potential customers change their quantity decision in direct relation to price. Similarly if competition has a lower production cost and/or is willing to accept a lower profit per unit of sale, the company may lose an order because the customer can obtain the same item at a lower cost from another supplier. If, however, the majority of companies within an industry have a similar cost structure and profit objective, the 'cost plus profit' approach can be a very effective pricing mechanism.

Demand orientated pricing
This approach is used by companies who consider that price is only one of the variables of the marketing mix to be utilized in the management of demand. Usually such organizations only consider price as a reference point for customers who have differing quality standards as a component of their purchase requirements. For this pricing concept to be applicable it is necessary that the market will respond to other variables such as promotional activity and a range of segments exist which have varying quality standards. In this situation, a manufacturer of breaded fillets may market a high priced species supported by advertising and a second lower

cost species which is supported mainly through emphasis on low price.

Competition orientated pricing

In the majority of market segments there is typically one company which leads in the initiation of price changes. A common solution in these circumstances is for the other companies to ignore concern over 'cost plus' or 'demand' pricing and to establish a price position in relation to the lead company. When the latter organization revises prices, the remaining companies follow in direct proportion to the magnitude of the leader's revision.

One form of competition pricing is found in the raw fillet sector of the fish industry where products are considered as relatively homogeneous. No one firm is capable of establishing a price significantly higher than the accepted market rate without risking a total loss of business. The producers and sellers are collectively understanding of the accepted 'price norm' relative to product availability. At the point where a change is necessary, this is usually initiated by one of the largest producers who provides a significant proportion of total catch. This move is the stimulus for the remaining companies to revise their prices.

In a market where the products may be more readily differentiated, companies are able to establish their own distinctive price policy because other factors such as quality are involved in the purchase decision. Nevertheless even in this situation, if supply conditions dictate an overall price reduction, it would be extremely risky for any single producer to ignore this factor, assume other elements in the marketing mix will sustain demand and not reduce prices.

Reaction to changing prices

Once a company has established a pricing policy in the fish industry, greater attention will usually be given to reacting to competitive pricing than to reviewing opportunities to initiate change contrary to overall market trends. Rapid fluctuations in the availability of raw material is a characteristic of the sea going section, reflecting both the seasonality of catch and the effects of environmental circumstances, eg bad weather, on catch per unit of effort. Aquaculture provides a more stable supply source because the farm operators can revise their volume of production decision in relation to the overall market situation. Nevertheless with few exceptions, eg trout, mussels and possibly marine shrimp by the mid'

1990s, this sector has yet to reach the point where output represents the major supply source and can, therefore, improve the stability of raw material to the point where pricing volatility is no longer a market characteristic.

Within the fish industry, market entry by a competitor exploiting a new resource is typically accompanied by the use of prices significantly lower than prevailing levels. Two examples are the launch of Japanese trout into the USA market in the early seventies, and the more recent attempt of Canadian fish companies to expand sales within the EEC. In certain instances the poorer quality of the new supply source would justify a lower price but more frequently the product is equal to that from existing suppliers, who might then be forced to reduce their own prices to a level below break-even. Eventually the financial burden can cause plant closures or severe cutbacks in the size of operating fleets, as, *eg* in the sardine industry in the UK which has been under continual price pressure from cheaper imports during most of the twentieth century.

On examining the long-term implications for the new competitor, it will be apparent that business expansion through price competition is a high risk proposition. The rationale for this strategy is that the new resource being exploited will yield a better catch per unit effort than in established fisheries, giving lower operating costs. Further cost benefits often accrue because the new fishery is located in the more underdeveloped areas of the world and labour is cheap. Many fish farmers apply the same logic in their pricing decisions. Having developed a new technology which offers a lower production cost than existing systems, they use the savings to enter the market on a price platform. Over time, however, the new fishery will exhibit a decline in catch, labour rates will rise, and in aquaculture another organization will eventually introduce a more cost-effective culture system. When this occurs, the company which built business on a price claim will in turn face a new market entry from a competitor offering lower prices. At this point the company will begin to lose customers because its claim has been invalidated by changes in the market environment.

For this reason it is recommended that a company which can produce a lower cost product should not offer reduced prices. Instead it should invest the higher profit margin in the creation of a strong customer loyalty through such mechanisms as promotion and development of more unique products which can distinguish the company from competition. Evidence for the benefits of this approach are illustrated by certain of the larger frozen fish and seafood companies in both North America and Europe. These organizations which in the last twenty years invested in major promotional activity and new product innovations have created a high level of customer loyalty. Consequently their sales remain relatively unaffected when a new producer enters the market offering lower prices, whereas groups which have emphasized price as the key element in their product claims are immediately in financial difficulty. Furthermore, from examination of the long list of companies which have gone into bankruptcy over the same period, it would appear that the majority had originally built their business on some form of price claim.

In view of this situation, if a company encounters price competition, it is preferable not to revise price but to examine alternative marketing actions to respond to changing circumstances. The nature of response will vary depending on the magnitude of the price competition. If it is only of a trivial level, *ie* less than 2%, then for most existing customers the price savings they could make by switching suppliers will probably be offset by the additional administrative costs associated with a change in supply source. The reaction to this level of price competition, therefore, is to minimize customer dissent by ensuring satisfaction with current service levels. This can be done by reviewing sales call frequency, minimizing distribution errors, the rapid handling of customer complaints and possibly the occasional relaxation of payment terms.

A minor magnitude of price competition, *ie* 2—5%, will require a more positive reaction through an increased level of short-term promotional programmes: coupons good on next purchase, 'money off' allowances on purchases for a brief period, sales contests. These methods will not be sufficient in the face of major price competition, *ie* 5—10%. In these circumstances the company will be forced to examine more effective mechanisms to influence buyer behaviour, such as a significantly increased level of advertising to achieve greater customer awareness of product benefits.

Extreme price competition, *ie* over 10%, will necessitate reviewing the potential for repositioning the affected product in order to avoid direct confrontation. This may be achieved through a revised promotional programme concentrating on a non-price orientated segment, but in many cases the product will have to be reformulated in order to meet customer need in the chosen segment. Should the nature of the product vitiate a repositioning, the company will be forced to consider market withdrawal

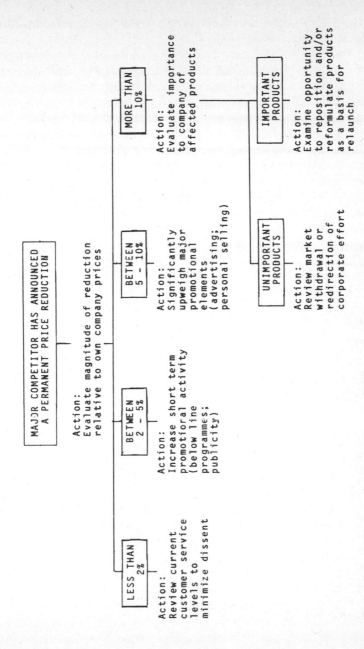

Fig 7.5   Decision paths in reaction to price reduction by competition

or redirection of corporate effort towards products less influenced by competitive pricing activity *(Fig 7.5)*.

Since the early sixties, a number of nations have recognized the economic benefit of expanding their fishing industry and utilizing the output to generate a favourable balance of payments through emphasis on export markets. To achieve this aim, such governments have been willing to subsidize the capital investment programme and fund market entry on a reduced price basis. Individual companies within the customer countries have frequently been forced out of business because they have been unable to obtain similar assistance from their own governments. Public sector recognition of the hazards of not protecting an industry from imports has, however, led to the introduction of tariff barriers in the fishing industry during the past decade.

Initiation of price changes

In order to initiate a price change, a number of factors have to be evaluated, the most important of which is the influence on sales volume. In theory this can be estimated through an understanding of price elasticity as described by the equation

$$\text{Elasticity} = \frac{(Q_1 - Q_0)/Q_0}{(P_1 - P_0)/P_0}$$

where Q_1 = quantity sold per period after price change
Q_0 = quantity sold per period before price change
P_1 = new price
P_0 = old price

Elasticity of one will mean sales rise (fall) by the same percentage as the price fall (rise). An elasticity greater than one means sales rise (fall) by more than the percentage price fall (rise). Less than one means a sales rise (fall) by less than the percentage price fall (rise).

In practice there are significant statistical problems in calculating elasticity with any degree of accuracy. A further complication is that demand is also influenced by the relative price of competitive protein sources such as meat or poultry. If, for example, meat prices are low, a major price increase for fish items will tend to further reduce sales. Alternatively, when meat prices are high, the same price change on fish will have less impact on sales volume.

A company should monitor the effect of price changes over a number of years, because this will provide at least a general understanding of elasticity. Without this knowledge, the marketing group would be unable to predict the outcome of a price change. This could result in an imbalance developing in other areas of the company such as matching production to sales, procuring adequate raw material supplies and the relationship between sales orders and finished goods inventory.

Elasticity is a reflection of the ultimate users' revision of consumption in reaction to price; but other factors are brought into play. One is the prevailing attitude amongst the intermediaries in the distribution system. For example, a situation may be encountered where consumers would accept a price increase but retail outlets have, for merchandising reasons, initiated a 'price freeze' and will not stock products unless suppliers guarantee a fixed price for a defined period.

Other factors which have to be assessed are the possible reactions to a price change by the government, the company's suppliers, and competition. The general trend in living costs is a political issue and hence a public sector group could decide to intervene if circumstances seem appropriate. Suppliers also monitor company product pricing, and an increase could trigger a revision of their raw material quotes. The company's own labour force constitutes another supplier, and it is not unusual for union negotiators to use pricing activity to justify a change in wage rates.

Probably the most important factor is competitive reaction. If competitors have a fixed price reaction policy (which can be determined by analysis of previous situations), the company has a relatively simple task. Where statistical patterns are less predictable, there is always the risk that competition may either not follow a price increase or perceive a price reduction as a threat and reduce their prices by an even greater amount. Because of the unpredictability of competitive behaviour, many companies use the mechanism of trade associations to provide the general consensus of opinion on price revision. This solution must be treated with care, however, because in certain countries there is legislation to ensure that trade association links do not develop into a formal structure through which companies can 'fix prices' across an entire industry.

8
Distribution

Channel structure

It is rare in today's society that the location of producer and the final customer permits direct trade between them. Typically they execute a transaction through one or more intermediaries ('middlemen' or wholesalers) who may be severely criticized by some primary producers who claim that wholesalers place excessive mark-ups on the product and thus reduce customer demand.

In reality, the location of processors relative to final markets, the perishability of the raw materials, the diversity of end-user needs and the seasonality of catch, all combine to form a complex set of variables which are costly to manage. Hence, rather than finding fault with the existing systems, the marketer in the industry should be cognizant of the issues involved and use the most timely and cost-effective method to ensure delivery of high quality products to the final customer.

The number of levels in the distribution chain will vary in relation to output and participant location. A small oyster farm marketing all production to restaurants in the same area would probably operate a two level system delivering direct to the end users. A larger farm whose final customers are not all located in the immediate area, may use a three level system with products being distributed by a wholesaler capable of servicing a larger geographic area. As output increases along with the distance between producer and final customer, more levels will usually be added to the system as additional intermediaries become involved. An example is the European system, where the fisherman sells to port wholesalers at auction, who in turn negotiate deliveries to inland wholesalers, who then supply retail outlets serving consumers.

Within the distribution system utilized there are five different types of flow — physical, title, payment, promotion and information — the nature of which reflects the responsibilities accepted by the various channel members. The role of channel members in a system for the marketing of breaded fillets (*Fig 8·1*) shows a six level system and if all the flows were superimposed on one diagram, this would emphasize the complexity of the distribution process. *Figure 8·1* also demonstrates there is a delegation of flow responsibilities by the primary producer and breaded fish processor. Such delegation is implicitly accompanied by a loss of control over the marketing of products by these two members of the channel. This will only occur if the companies concerned perceive some advantage by delegating certain roles to the intermediaries.

The most important advantage is the reduction in the number of transactions. If three customers required products from three producers, this would involve $3 \times 3 = 9$ transactions to consummate the purchase process. If an intermediary is added to the system, *eg* a retail outlet, the number of transactions is reduced to $3 + 3 = 6$. Furthermore in most channels, the output of a single producer is usually much greater than the needs of a single end user. To minimize distribution costs, the producer will ship in bulk quantities, *eg* a 30,000lb truckload of fillets packed ten 5lb boxes to a 50lb master carton, but individual end users may only wish to purchase a single 5lb box of the product. The intermediary will, therefore, accept the role of warehousing the truckload quantity and breaking down the loads into individual boxes for sale to the final customer.

Channel selection

The standard approach to channel choice is to select the distribution path most likely to ensure the availability of products at the purchase point utilized by the company's target customers. The optimum channel will be that which best deals with the constraints created by the characteristics of the product, the customer, the available intermediaries, competitive activity, company structure and the business environment.

Fresh fish and seafood are highly perishable and demand the use of the shortest possible distribution channel with the minimum involvement of intermediaries, to reduce the delays associated with repeated changes of ownership. In order to ensure quality retention at the time of consumption therefore, many companies market the product direct to the final customer, *eg* a fish company which establishes distribution outlets equipped with trucks in major markets and delivers the product on ice to restaurants. Where the location or financial resources of a primary producer preclude a direct marketing system the company will be forced

84

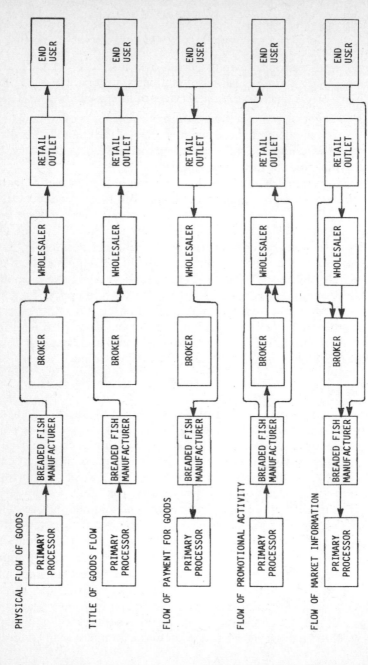

PHYSICAL FLOW OF GOODS

TITLE OF GOODS FLOW

FLOW OF PAYMENT FOR GOODS

FLOW OF PROMOTIONAL ACTIVITY

FLOW OF MARKET INFORMATION

Fig 8·1   Component flows in a six level distribution channel

to distribute the product in a form which offers a longer shelf-life, *ie* canned or frozen, unless the responsibility for rapid distribution can be passed to an intermediary.

Where the final customers are dispersed over a wide geographic area and purchase small quantities on a frequent basis, it is usually preferable for the primary producer to utilize the services of intermediaries within the various markets, *eg* wholesalers supplying retail outlets, to manage the distribution process. Alternatively if the primary producer markets to a small number of large customers, *eg* major restaurant chains, who place large orders on a less frequent basis, the company will be able to minimize involvement of intermediaries.

Although product form and customer buying behaviour may permit the usage of numerous intermediaries, it is necessary that the available middlemen are capable of executing the distribution responsibility. If, however, potential intermediaries have insufficient aptitude or resources to be responsible for the timely delivery of goods with no reduction in quality standards, the company may be forced to accept greater involvement in the distribution process. For example, in some countries there are no wholesalers equipped to handle frozen products. If a company wished to enter such markets, it would be necessary to invest in the establishment of a frozen food warehouse and also supply retailers with freezer cabinets.

When the majority of the industry uses a specific distribution channel and the final customer wishes to make product comparisons at the purchase point, then a single company will usually have to utilize the same distribution process to generate sales. This is not mandatory, for the company may decide that there is excessive competitive pressure in the existing channel and, in order to avoid this situation, attempt to locate an alternative system. One example is the smoked product sector of the Scottish salmon aquaculture industry where some processors decided to avoid the traditional wholesale/retail distribution system and market their product direct to private households using mail order.

A very common situation within a distribution channel is where the intermediaries are only willing to handle products from a limited number of suppliers. A characteristic of the fish industry is the presence of a large number of primary processors, and there is intense competition for entry into available distribution channels. In these circumstances the company with a limited product range and/or insufficient financial resources to support the range with major marketing activity will be in a weak negotiating situation when endeavouring to persuade a middleman, *eg* a food service distributor, to distribute the company's output.

A further constraint influencing available channels is imposed by the characteristics of the business environment. Economic conditions could increase the importance of price in the customers' buying behaviour, causing channels providing high service levels — and higher costs — to be less attractive. Government regulations (enforceable in a Court of Law) defining acceptable bacteria levels might create a further obstacle in the channel selection process.

The company will usually face a range of alternative channel choices some of which will be vitiated by the influence of the previously described constraints, with the final choice being made in most cases upon the basis of two criteria: economic cost and availability of management control mechanisms. Of the two, the economic issue tends to be the more important, with the objective being to select the channel system which provides an optimum balance between sales volume and distribution costs.

A simple illustration of this concept would be a processor who is entering a new market supplying retail outlets and evaluating the choice between opening a sales office or employing a sales agent. The use of a company sales force has the advantage that the attention of the sales-person is solely concentrated on the company's products and they have the expertise to advise customers on optimum purchase decisions. An agent will typically represent a range of very diverse products and will lack in-depth knowledge of the company.

Clearly a key issue to determine is which of the alternatives is likely to generate the greatest sales volume, estimated by drawing on previous corporate experience and, if necessary, outside opinion of similar situations. Once the probable sales levels have been evaluated, it is necessary to calculate the costs for the two choices. Agencies typically operate on a fixed percentage of sales and hence a lack of sales will not cause any distribution costs to be incurred. A company sales office, however, represents a high level of fixed costs due to such factors as establishing the facility and the staff salaries. Because of this cost variance and the possible difference in sales, one can best evaluate the alternatives on a comparative return on investment, ROI, calculation where

$$ROI_i = \frac{S_i - DC_i}{DC_i}$$

where ROI_i = return on alternative i
   S_i = estimated sales revenue for alternative i
   DC_i = estimated distribution costs for alternative i.

The issue of control availability within a channel depends on whether the intermediaries will accept recommendations from the producer company on quality control, merchandising programmes and pricing. If, for example, the company feels that quality control is vital to the acceptance of the product by the final customer, then when two channels offer a similar ROI, the tendency would be to choose the distribution mechanism most responsive to guidance on product quality.

Channel review

Distribution channels are continually undergoing change. It is a key responsibility of the marketing group to be aware of this situation and react to change before it can adversely affect sales volume. A company may have decided to focus on marketing non-branded bulk packs to a specific chain of retail stores. If after a period of success this intermediary ceases to achieve an optimum performance because of competitive pressures, this development should be identified and a more growth orientated intermediary sought.

Change does not just occur in subcomponents of a channel, but can also occur across a total channel system. For example, in both North America and Europe, the fifties and sixties saw a decline in fresh fish and seafood sales as many of the intermediaries de-emphasized this business segment to concentrate on frozen products. In the past five years there has been, for various reasons, a resurgence in the market for fresh products but only the more innovative companies in the industry recognized this growth early enough to capitalize on it. These innovators are now in a position to dominate the fresh fish segment and have some capability to bar entry by other suppliers who have also become aware of this trend.

Physical distribution

The purpose of physical distribution is to ensure delivery of output to the point of sale in a timely and cost-effective fashion. In the fish industry, the perishability of raw material, seasonality of catch and distance between production and consumption combine to cause physical distribution to be a major area of responsibility.

The readily identifiable cost components of a system include:
(a) moving goods from point A to point B
(b) storage
(c) funding the acquisition and holding of inventory.

To this must be added a less quantifiable element, namely the cost to the company of sales lost because customers will not accept delivery delays and do not purchase products. The selection of the appropriate physical distribution system process is similar to that for the channel choice. The company will face a number of choices and selection will be based upon balancing total sales against an objective of minimizing delivery costs.

The simplest system is a single plant, single market situation. In the fish industry the catch is rarely sold in an unmodified form except for certain types of shellfish, eg mussels, lobsters. Many finfish are headed and gutted, and the most extreme case, fillet production, results in the discarding of 60—75% of original landed weight of the product. As there is little point in paying shipment costs on material which will not be included in the final sale, the usual solution is to locate the processing plant near to the point of landing rather than the final market. Where significant further processing occurs, eg the production of breaded fillets, the plants are often located nearer the final market due to the high cost of transporting low density, high volume materials such as batters and breadings over long distances. This latter situation is, however, rarely resolved on distribution costs alone, as the cost factors of the processing operation also have to be taken into account. Labour and land costs tend to be higher in urban environments and savings in distribution could be more than offset by higher processing costs for a plant located within urban markets.

The alternative modes of delivery available today represent significant differences in the time taken, with the costs varying in direct proportion to the delivery period (*Fig 8·2*). As the delivery period lengthens, the cost of lost sales increases, due both to the unwillingness of customers to accept delays and to the deterioration of product quality to a level unacceptable to the final customer. The combined costs of delivery and lost sales can be used to evaluate the minimum total distribution cost and define the optimum delivery system (*Fig 8·3*). It is apparent from the curves for the three example product forms (fresh, frozen and canned salmon) that time, because of the effects on product quality, is a major factor influencing the cost of lost sales. Hence for the short life fresh product, air freight or the more rapid forms of road transport are the optimum methods of distri-

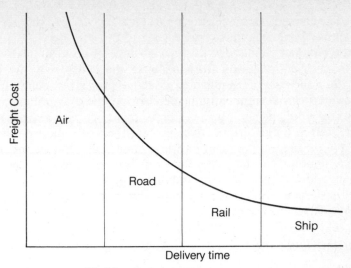

Fig 8·2   Freight costs and delivery time

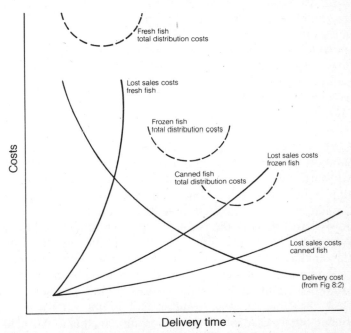

Fig 8·3   Evaluation of total distribution costs for three forms of salmon

bution. The canned items have the longest product life and this permits the use of slower but lower cost systems such as rail or water.

As the company expands into a wide range of geographically diverse markets, distribution costs from a single manufacturing site may be greater than the cost of establishing a second processing centre nearer to the final market, but frequently the production and distribution costs in a free market would be against the establishment of a new plant in the final market. Action by governments to protect their own industry has often resulted in the creation of tariff barriers, *eg* the duty on fish imports from the USA imposed by the European Economic Community. When such tariffs are added to the operating and distribution costs, this could result in the establishment of a secondary plant inside the tariff barrier being an economically viable concept.

Inventory decisions

Acquisition and holding of stock is a short term investment decision usually funded from a company's working capital account. The fishing industry usually faces seasonal production but year round demand. This will necessitate holding inventory for a longer period than in most other industries and an accepted standard for many producers is to assume an average inventory during the year equal to four to five months sales. Most companies are therefore forced to borrow funds from a bank to meet this cost and consequently the interest charges on inventory loans are a major expense element.

Market recognition of the seasonality of output is reflected in price trends at their lowest during the catch period and rising gradually over the year as shortages develop. The issue facing the primary processor is to forecast when the difference between product cost and market price is at the point to maximize profit. Unfortunately, too many companies do not reflect interest charges in their inventory valuation and hence delude themselves into believing the point of highest market price is the time to sell. When in fact if one adds the cumulative expenses of holding inventory, achievement of maximum profit may occur by entering the market at an earlier point in the year. This decision is further complicated by the unpredictability of actual future price trends. The 'sell or hold' issue is critical; therefore in most companies it is not left to the marketing department but is controlled by senior management. An error in this matter can have a major impact on overall financial performance and indeed is probably the most frequent cause of bankruptcy in the industry.

Although members of the retail food service markets recognize the seasonal nature of output for fishing and aquaculture, they nevertheless expect to be supplied on a year round basis. To achieve this, suppliers of canned or frozen products would have to accept the concept of 'a 100% service level', never permitting the inventory level to fall to that at which shortages occur. Unfortunately, it is a fact that there is an exponential relationship between inventory costs and service levels and this should cause the marketing department to evaluate the level of service that the company should be willing to provide. Even where gross profit per unit sale is linear, the exponential nature of the inventory cost curve will mean that as customer service levels approach 100%, net profitability, *ie* gross profit minus inventory costs, will decline (*Fig 8·4*). One solution is to establish an ROI objective calculated from the formula:

$$\text{ROI} = \frac{\text{Gross profit}}{\text{Sales}} \times \frac{\text{Sales}}{\text{Investment in inventory to sustain sales}}$$

and by iterative means establish the sales level which will deliver the required financial objective. In most situations, this will result in the supplier company operating a service level somewhere below 100% and customers being forced to accept that occasionally producers will be out of stock on certain items during the year. An additional complication facing companies involved in the manufacture of a range of processed fish or seafoods is the balancing of processing costs against inventory carrying charges. On the manufacturing side of the equation average production costs decline the greater the volume produced because set-up costs are minimized and plant workers have sufficient time to optimize processing efficiencies. However, longer production runs will generate finished goods in excess of immediate customer need and inventory costs will begin to rise. The cumulative costs per unit for these factors will usually combine to make it possible to evaluate graphically an optimum order point to minimize total product costs (*Fig 8·5*).

In the past few years interest charges on loans and the cost of energy expended on distribution have been rising at a rate in excess of average inflation. Traditionally, marketing personnel have placed little emphasis on the distribution responsibility on the assumption that this is a minor factor in overall responsibilities. For many companies, however, total distribution costs now represent over 20% of total operative expenditure,

91

and some predictions show this could rise to over 40%. It is, therefore, vital that the marketer begins to place greater emphasis on distribution during the management of the total marketing mix.

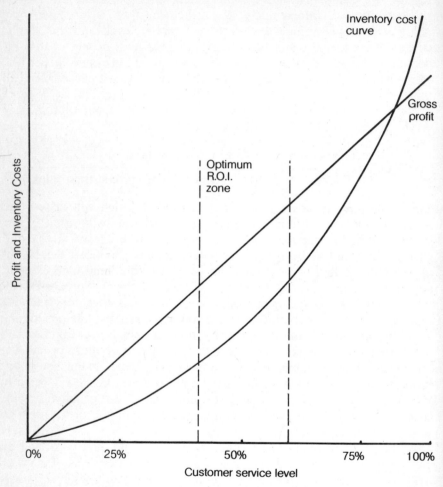

Fig 8·4   Service level, profit and inventory costs

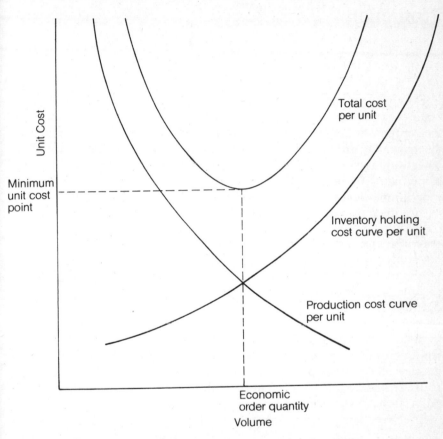

Fig 8·5   Graphical evaluation of economic order quantity

9
Promotion

Correctly positioned product, competitive pricing and effective distribution will not result in sales unless potential customers are made aware of the comparative advantages of the company's product. It is necessary for the marketing group to develop an integrated system of customer communication, and this is achieved through promotional activity. The four components of promotion are advertising, personal selling, below-line programmes and public relations. Of these, advertising and personal selling are the most important in building long-term customer loyalty.

Below-line programmes such as free samples, coupons 'good on next purchase', price reductions for intermediaries and sales contests, usually have only a short-term impact on sales. Sales will rise during the period the programme is in effect, but upon its termination sales will fall and eventually stabilize at a level similar to that prior to programme introduction. A major aspect of public relations is the mention of specific elements of a company's operation by the media, and is distinguished from advertising in that the information is published free of charge. Although publicity can assist in creating or reinforcing the company's image in the market place, the company has no control over the media message and hence the mechanism is of limited assistance in the overall promotional programme.

Effective management of the information given to customers and intermediaries is achieved through advertising and/or personal selling. It is rare for a company to use only one of these two choices, since it will usually have a mixed promotional plan with emphasis related to the market structure, balanced with the objective of minimizing the cost of effective communication of the promotional message.

Any communication to the customer is more effective if presented in the form of a dialogue in which the promoter imparts information and is

94

able also to respond to requests for further details. In theory, therefore, the optimum communication mechanism in marketing a product would be a sales force calling on every potential customer. This is quite feasible where a few customers are purchasing large quantities of products and are located so that the sales force expend minimal time travelling between sales calls. As unit purchase per customer declines and dispersion of customer location increases, the cost of personal selling will rise. A point will eventually be reached where advertising — which has the drawback of being a one-way communication process — will be more cost-effective in the communication of information (*Fig 9·1*). As a result, greater

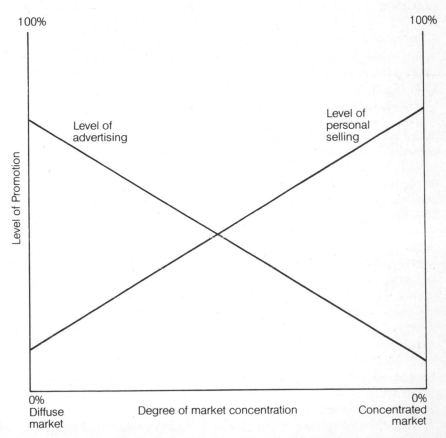

Fig 9·1   Market structure and promotional activity

emphasis is placed on personal selling in the industrial market sector and the situation reversed in consumer markets with advertising becoming the major promotion mechanism.

The balance between advertising and personal selling in a market is rarely steady because customer behaviour and channel structure are continually undergoing change. This effect is often apparent with a new retail product. Early in the life cycle, intermediaries will require frequent visits from the introducing company in order to communicate product features, marketing plans and merchandising techniques. Concurrently, advertising is utilized to generate awareness amongst potential customers. Once a product is established and moves into the maturity stage, minimal sales effort will be necessary because the intermediaries will have become fully conversant with the product. Advertising will continue, however, to sustain consumer awareness for the brand.

Theoretically it is quite simple to evaluate the optimum balance between advertising and personal selling relative to sales. Given a fixed total promotional budget, an analysis of sales volume for the various combinations of personal selling and advertising effort will permit identification of the point of maximum sales. In practice the acquisition of the data for such analysis is made extremely difficult because of the changing nature of the market environment. Hence the majority of companies utilize observations on purchase behaviour and geographic dispersion to produce a generalized conclusion on the balance of advertising to personal selling. In the fish industry this will usually result in companies emphasizing advertising in retail markets; and in the food service (or industrial) sector relying mainly on a sales force accompanied by a minimal weight advertising campaign to sustain brand awareness or communicate general information on some of the company's more important products. What is much rarer is for such companies to refine further their promotional mix by controlled test markets to examine the optimum balance of the components of the promotional mix.

Advertising

The four components in the simplest form of advertising communication model are:
(a) the 'communicator' or source of information
(b) the 'channel' or route through which the information is passed
(c) the 'message' or specific information communicated

(*d*) the 'audience' or group of potential customers to whom the message is directed.

A group such as the Food and Agriculture Organization (FAO) may wish to build consumption of fish as a protein source in a developing country. The FAO is the communicator and the message could deal with the issue of fish being a tasteful, economic ingredient for main meals for the whole family. In this instance the audience will be the majority of consumer households in the country. As the requirement is to reach such a large segment of the population an appropriate channel of mass communication such as radio would be chosen.

It should be recognized that any advertising will not, without the other components of the marketing mix, result in increased sales. All advertising can achieve is to increase awareness amongst potential customers about the benefits of product usage to the point where the buyers are most positively disposed towards purchase. At this point factors such as other elements of the promotional mix, the product and the price relative to alternative offerings, will all be considered by the customer. If these factors provide further positive reinforcement of the original information provided by the advertising then the customer will purchase the product.

The first point in the advertising planning process is to decide on the content of the message from the communicator. Development of an appropriate message is a highly skilled task and most companies will employ the services of an advertising agency to manage this process. The most effective messages are those which focus on a single product benefit and comunicate this in a believable and desirable format. The earlier scction on product positioning and market research revealed that customers perceive most products as offering a multi-faceted set of benefits based upon a rangc of attributes. It is necessary to use available information on customer behaviour, *eg* the type of material generated during the case example on canned salmon, to select the most important benefit element in the purchase decision process. Benefit emphasis might relate to creating greater awareness of a specific product feature, which is often the situation for new products, or in the case of the growth/maturity phase of the life cycle, focus on sustaining a claim of superiority over competitive offerings. A recommended approach in the development of an advertising message is to specify the chosen claim in the form of a creative strategy which states the product benefit and the characteristics or attributes of the product to be used in justifying the selected benefit.

Implicit in the development of the creative strategy is a decision on the

97

audience to be reached as it is from this source that information on which to base the strategy is drawn. For most companies this audience will be the same as that identified during the decision process on product positioning. The major cost involved in mounting an advertising campaign may, however, cause the company to decide it cannot afford to reach all the potential customers in the market. A common solution is to focus advertising on those customer groups which, it is felt, represent the most probable source of medium/heavy product purchase. For example, it would be desirable in the previously mentioned FAO campaign to communicate to all households in the country but this would be unaffordable. Hence the selected advertising target might be households containing adults in the age group 21—45 with children, because this group offers the highest potential level of consumption per family unit.

Although advertising can represent the greatest single item of expenditure of the total marketing budget, few companies apply more than a minimal effort to planning how such funds can be most effectively utilized. The main reason for this is that only by interacting with other marketing variables can advertising result in the generation of increased sales. Thus accurate measurement of the effectiveness of this area of promotional activity is difficult.

In recent years, it has been established that, assuming all other factors remain constant, the relationship between sales rates and advertising is typified by a concave or S-shaped curve (*Fig 9.2*). The important features of these theoretical curves is that (*a*) there is a threshold level below which no increased sales will occur, and (*b*) beyond a certain point there will be no further improvement in total sales irrespective of any increase in advertising expenditure. Despite this theoretical model, the majority of companies still apply methods such as:

(*a*) Affordability — where the budget is set on the basis of what the management believe can be afforded.
(*b*) Percentage of sales — where the advertising budget is calculated on the basis of fixed proportion of forecasted sales.
(*c*) Competitive parity — where the company attempts to relate expenditure to a specific proportion of competitive spending, *eg* 50% of competition; equal to competition; 150% of competition.

These approaches can all be criticized on the grounds that the specified budget could be at a point on the sales/advertising curve where (*a*) a small increment could significantly increase sales, (*b*) sales have already

98

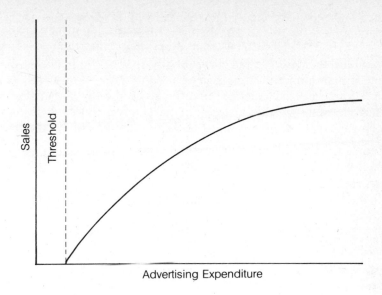

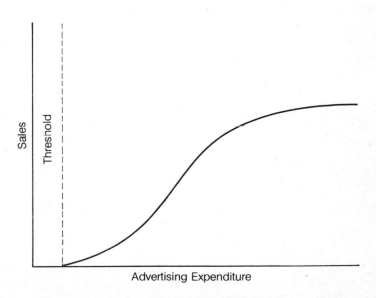

Fig 9.2   Sales advertising expenditure curves

plateaued and expenditure could be reduced with no decrease in sales, or (c) expenditure is below the threshold and would have no influence on sales.

To establish the nature of the curve for any specific product will usually require test marketing to estimate sales at various spending levels. These data can then be utilized to define the variables for the equation to describe the sales/expenditure relationship. An alternative approach is to use management experience by posing a series of questions such as:

(1) What is the current market share and advertising expenditure?
(2) What would happen to market share if advertising expenditure was increased by 50%?
(3) What would happen to market share if advertising expenditure was reduced by 50%?
(4) What would happen to market share if all advertising activity was terminated?

These four values can then be used to construct a directional indication of the sales/advertising curve.

An accepted theory on the communication/learning process is that the provision of information results in the potential customer moving from awareness to interest and then to evaluation. At this latter point other factors, *eg* the availability of product at point of purchase and below-line promotional programmes, have influence and the customer will move from evaluation to purchase. If through experience or experiment the relationship between awareness and purchase can be established, this communication model can be used to decide upon the required advertising budget. The marketing group assigned to the FAO project may decide their objective is to achieve a 20% trial rate in the country which contains 20 million households of the audience in the age group 21—45 with children. Their previous experience indicates that (*a*) on average each potential customer has to be exposed to the advertisement three times to register the message effectively, and (*b*) the conversion rate from awareness to trial is in the region of 4:1. The 20% trial rate objective represents purchases by 4 million households, *ie* $0.20 \times$ total households, and to achieve this 4 million $\times$ 4 (or 16 million) households would have to be exposed to the campaign. Given this coverage, which equates to 80% of the target audience, and knowing the total number of advertisements required for an average exposure of three messages per household, the advertising budget can be calculated.

A financial solution for the above example could only be developed if information on costs and media vehicle effectiveness were available. Media vehicles are of two types, print and audio-visual, with the latter encompassing radio, television and cinema. In many countries radio provides an excellent mechanism for communication to the majority of the population. Commercials can be developed at low cost, and the lead time from concept to finished advertisement can be very short. Television and cinemas offer a distinct advantage over radio in that the audio message is combined with a visual element, probably the most effective mechanism of communication. Television is an expensive technology both in terms of transmission and reception. Hence only in the more affluent countries does one find it is a mass media vehicle. In less developed nations the majority of the population is only able to afford occasional visits to the cinema and certainly does not have access to a television set.

Audio-visual communication is ephemeral whereas print has the advantage of being more enduring, with the advertisement often being read more than once by one or more potential customers. Newspapers are probably the print vehicle which can deliver the larger audience to the advertiser. Magazines tend to have a smaller circulation and usually each magazine has a very specific reader profile. Further, because of the printing processes used for most magazines, they provide a much higher reprographic quality. Outside print media such as billboards or posters are considered a very secondary medium in the Western World. However, in countries of low literacy rates and insufficient *per capita* income to permit widespread purchase of newspapers or magazines, the outdoor media can become the most effective print vehicle for broadscale market communication.

It is apparent that the diversity of media vehicles within a country and the variance in their ability to reach audience types depending on overall living standards, causes choice of media to be a complex subject. Thus most companies choose to make use of the expertise available within advertising agencies to plan and execute advertising campaigns.

The 'affordability' constraint

The cost of advertising reflects the supply of media vehicle alternatives and the demand for space. Relative to many industries, absolute sales and profit as a percentage of sales in the fish industry are quite low. This means that fish companies rarely have sufficient buying power in the form of controllable marketing funds to purchase more than a minimal weight of advertising. Hence a frequent dilemma facing the marketer is how to obtain an improved sales performance with limited financial resources.

Typically there is an inverse relationship between the level of further processing in the industry, which provides an opportunity to create 'added value' and profitability as a percentage of sales. In the USA shrimp industry this is reflected in the advertising for the various product forms. Shell-on shrimp is a commodity orientated item with heavy price competition between suppliers. Profitability is improved by peeling or cooking, but the highest profitability is achieved in the processed breaded shrimp segment. The advertising campaigns of USA breaded shrimp companies use television or magazines and newspapers as media vehicles. The peeled product is rarely advertised on television or in magazines because of the high costs, and campaigns are limited to regional newspapers where space costs are lower. Shell-on shrimp tends not to be advertised except for an occasional mention in trade magazines.

Even the largest companies in the industry face the problem that sustained television or magazine advertising across the entire product line is a prohibitively expensive proposition. The usual solution is to choose one product type, either an important new item or a sales category leader, as the core product around which to centre advertising activity. It is assumed by these companies that this activity will increase overall brand awareness and indirectly assist sales rates for other items. In reality, although heightened brand awareness is of benefit, significantly higher sales rates for other items cannot be achieved unless the company can afford advertising specific to such items.

The industry in many countries is characterized by being composed of numerous small companies. These commercial entities face problems in funding advertising unless they are willing to forgo current profit, invest in a major campaign and assume repayment from a future improvement in sales. In an industry where fluctuating landings create a highly volatile sales trend, this is a high risk proposition. One alternative is to use national sales revenue to support geographically restricted advertising in selected areas of the market. The areas chosen will tend to reflect the overall corporate marketing objective. If the company is in an expansion phase, advertising could be directed towards geographic areas which offer the maximum opportunity for sales increases. A more controlled growth strategy might be reflected by advertising in the company's stronger markets to assist in both generating incremental sales and defending the product line against market entry by competition.

Where smaller companies accept that effort to stimulate improvements in the overall generic sales rate for a product form is of benefit to all parties, an adequate campaign can be funded by a pooling of advertising budgets. This frequently occurs where the companies are members of an active trade association. Salmonoid farmers in both Europe and North America have used this method to build demand for trout, their argument being that increased customer awareness for the industry will be of benefit to all participants in the marketing of the product.

Such programmes may fail if members disagree on the specific nature of the advertising message and, to avoid dissent, a compromise campaign may be developed which attempts to encompass all opinion. The consequent advertising is often confused and thereby fails to communicate effectively a specific, single-minded benefit claim.

The 'pooled resources' approach can also be achieved by a company and an intermediary. The commonest form of promotion is that of a retail outlet which links the company product to other items carried in-store as a mechanism to generate additional store traffic. If a company decides this is an appropriate mechanism, it should ensure that the intermediary (a) shares the same overall objective and (b) is willing to provide the company with adequate representation in the advertisement. For example, a company marketing premium quality fresh seafoods would not be wise to link up with a supermarket group which emphasizes low prices and has a customer type biased towards the lower end of the social scale.

Some governments are willing to provide support to the fishing industry. Traditionally this has often been in the form of subsidies for boat operators and loans to assist in the construction of new processing facilities. It is becoming increasingly apparent, however, that if no action is taken to create demand for the expanded output facilitated by public sector financial intervention, landed fish prices simply trend downwards as catch increases. To overcome this problem, certain governments are redirecting their efforts towards assisting industry to build greater customer awareness through advertising the benefits of fish and seafoods. Hence for small companies unable to support their own promotional programme, one solution is to approach the government on a unified front to request diversion of funds away from expansion of output into building new markets through promotional activity.

The traditional fish industry is facing rising operating costs and declining profit margins which is further compounding the problems of how to fund advertising. In contrast, advances in aquaculture technology are reducing production costs and providing opportunity for high profit margins. To date most aquaculture groups appear to be using incremental profits to fund output expansion through further capital investment or attempting to heighten share values by paying high dividends. Others are merely using the cost advantage to support a penetration pricing strategy to build market volume. The strategy which is overlooked is that of retaining a premium price position (which for most culture products is sustained by a quality superiority over equivalent products from natural populations) and utilizing the high margin to fund an advertising campaign to build customer loyalty for the company's product. It is a proven fact in the food industry (as demonstrated by companies such as Kraft, General Foods and Nestlé) that this is the only approach to create a dominant and profitable business position over the long term.

Failure to recognize this latter path by existing companies in aquaculture could over time be the cause of their downfall. At the moment few of the major marketing orientated companies have more than a marginal involvement in aquaculture. A large number are, however, now actively monitoring current trends to identify an appropriate entry strategy. Once they move into the arena, their first action will probably be to re-invest profit from culture output into the type of advertising campaigns which form a foundation stone of their existing product ranges. Should this occur and the smaller companies have failed to establish any degree of customer loyalty, it may not be possible for them to generate sales on any other basis than price competition. The accompanying implication of this action would be one of very low profits per unit of sale.

10
Personal selling

There is probably no individual in business more maligned than the salesperson, who is often pictured as spending large sums of money entertaining customers to 'buy orders'. In reality a salesperson is a professional who carries the major responsibility of generating the sales revenue upon which the company depends for survival. This is especially true in the fish industry where the role often represents the major promotional vehicle.

It is rarely possible to generalize on the objectives of a sales force. Their role and activities are greatly influenced by variables which include market structure, product complexity, distribution channels and customer location. This can be demonstrated by examining the multifaceted role of a sales director responsible for developing business for a new species of fish which a company believes has potential to be marketed in the block form for use by processors manufacturing breaded fish portions.

At the outset the marketing group will seek the Sales Director's advice on possible product positioning and competitive activity in the market. The director would then identify potential customers and on that basis would develop the sales presentation. At this point, the director begins to plan activities outside the company, by arranging appointments and making sales presentations. Hopefully, the meetings with customers will eventually result in orders, but the role of 'order taker' is only one component of the sales management responsibility. The new species will probably have features different from existing items and the director may need to be involved in technical discussions with the customer to solve any potential processing problems. Maximum customer satisfaction is a vital sales objective and hence the director will endeavour to ensure that no problems occur over payment and delivery terms. Finally, the director will also need to avoid future problems due to post-purchase dissatisfaction and would be advised to spend time with the customer production or quality control personnel to reassure all parties concerned that the new product is meeting expectation.

The above example provides a situation where the salesperson's tasks encompassed internal adviser, definer of customer profiles, locating customers, developing presentation, order taking, as well as overseeing distribution, payment and customer post-purchase satisfaction. These roles will vary in importance and some sales personnel will be involved in only one or a few of the components of the total range of activity. A driver of a truck delivering fish to retail stores might, for example, only be involved in order taking and delivery. A regional Sales Manager may have agents who are responsible for order taking and administration, while he concentrates on defining potential customers, developing presentations, training agents in actually executing the presentations, and managing the control system designed to evaluate agent performance.

A major influence on the modern sales role has been the recognition that the salesperson is not involved in a seller-buyer relationship simply to generate orders, but is actually a member of a partnership in which both seller and buyer share the common objective of optimizing the buyer's purchasing and resale activities. The salesperson who recognizes that he has a wealth of knowledge about his company's products and can thus assist the buyer in making more effective purchase decisions, is usually the individual who will, over the long term, maximize sales revenue for the employer. This approach is now referred to as Total Systems Selling in the industrial sector, with the salesperson analysing the customer's business as the basis of a formal recommendation on how his company's products can be of assistance. The equivalent approach in the retail sector is known as Programmed Merchandising and the concept is compared with the more traditional approach in *Table 10·1*.

Because of this diversity of responsibility in selling, it is mandatory for the marketing group clearly to define their overall planning objectives and to reach agreement with the sales force on how these can be translated into specific sales tasks. For example a company marketing canned crab and introducing a new pack size may decide that the order of priority is (*a*) the existing product in existing accounts, (*b*) introducing the new pack, and (*c*) to increase overall retail distribution. This decision could be translated into the following definition of tasks and time allocation for the sales force:

Activity	% of time
Existing accounts/established pack sizes	60
Existing accounts/new pack size	20

New accounts/established pack sizes 15
New accounts/new pack size 5
100

Table 10·1 THE TRADITIONAL SELLING ACTIVITY BETWEEN PROCESSOR AND
RETAILER COMPARED TO PROGRAMMED MERCHANDISING

Issue	Traditional Selling	Programmed Merchandising
Location of sales call	Individual retail outlets	At head office of retail chain
Timing of call	Route pattern/call frequency decided by processor	Timing and frequency defined by retail chain buyer
Information reviewed	Processor price quotes and product availability	Merchandising and promotional plans of processor and retailer
Processor participants	Van salesman	National accounts sales team
Retailer participants	Store manager	HO chain fish buyer, food products merchandiser
Processor goals	Large order	Sustaining long-term relationship with the retail chain
Retailer goals	Next period sales, low inventory	Specific near-term chain promotional objectives, *eg* increased profit per square foot, higher overall store traffic, new store openings
Mutual goals	None	Sustained growth in terms of profitability and investment

Sales force design

In most companies there is a proportionate relationship between total sales and the size of the sales force. The cost of a salesperson is usually higher than the average cost per head for other employees of a company. This is because salespersons can usually command a salary which reflects the importance of their role to which has to be added the travel costs incurred visiting customers. The location of the territory some distance from home office may also require the opening of a local centre to provide an administrative base for the sales personnel. In view of this situation, a careful assessment has to be made to balance the cost of the sales force size against sales revenue objectives.

One approach is to define workload in terms of desired call frequency, estimate the average call rate capability of an individual salesperson, and from these data estimate the required number of salespersons. For example, an aquaculture group, which produces pen raised salmon sold in both fresh and frozen form, markets output in both the retail and food service sector. An evaluation of customers indicates the following level of required call frequencies:

Call Frequency	Calls per Year	No of Accounts	Total calls per year*
Once a week	52	14	728
Once every two weeks	26	46	1196
Once a month	12	78	936
Once every two months	6	112	672
Once every four months	3	210	630
Once every six months	2	308	616
Total calls	N/A	768	4778

*Calculated from calls per year × number of accounts

The wide geographic territories and the type of administrative responsibilities involved in co-ordinating fresh deliveries result in salespersons being able to spend only 50% of their time per week in actual account presentations. This is reflected in an average call level of 12 per week, and given an average work year of 44 weeks (the balance being consumed by holiday and time at head office meetings) this represents a call rate of

$$12 \times 44 = 528 \text{ calls per year.}$$

Given the total overall company call requirement, the sales force size is equal to

$$4778 \div 528 = 9 \text{ salespersons.}$$

There are a number of structures available on which to base a sales organization and final choice will depend upon which is the most effective to ensure achievement of defined objectives. The simplest method is to place each salesperson in a defined territory, responsible for the company's entire product range. This approach works well for homogeneous products and hence is very common in the fish industry. Where specialist knowledge of the product is required of the salesperson, or the product line is highly diversified, the company will assign individuals to specific products. This is a rarer situation in the industry but companies which trade various forms, *eg* commodities, export items and processed products, across a broad geographic area relying mainly on telephone or telex communication to customers (sometimes assisted by agents at local market level) may use this structure. A disadvantage of the approach is that more than one individual from the company may be interfacing with the same customer or travelling to the same place. This represents a duplication of effort, but on the other hand it is assumed the greater product expertise of each salesperson will result in better customer service and thus greater sales volume.

In order to solve customer problems, the salesperson will require understanding of the type of need specific to each source of sales. Where there is a diversity of problems, one solution is to organize the sales force to reflect customer structure in the market. This is a relatively common approach in the fish industry where a sales force could, for example, divide the food servicing group into three areas: major restaurant chains, wholesalers servicing individual restaurants, and public sector catering operations, *eg* school and hospital canteens.

As the complexity of the sales activity increases, it may be necessary to combine one or more of the above structures to achieve the most efficient sales operation. The sales force may have originally been organized on a territorial basis which then evolves into a territory-product structure, *eg* within each geographic area individuals are responsible for food service or retail products. Further evaluation can result in a territory-product customer structure, *eg* the above food service personnel begin to specialize in selected types of food service outlet. It must therefore be recognized by management that no structure, however effective upon original introduction, can be expected to remain as an optimum solution if change occurs in the company and/or customer activities. Hence a periodic review of sales force responsibility should occur during which the structure issue should be a major area of emphasis.

Even a perfect structure is of no benefit if it contains incompetent or poorly motivated individuals. Given the need for a successful sales operation, major attention must therefore be given to the recruitment and management of the sales force. The actual hiring procedures may be the

109

responsibility of the sales department or delegated to the personnel division. The first step in the hiring programme is to draw up a specification showing the nature of the job and the desired education/ experience profile of potential applicants. More senior sales posts, *eg* a manager responsible for a territory containing four area managers, will usually be reflected in a requirement for a minimum number of years involvement in selling and/or specific statements of exposure to certain sales roles with previous employers.

It is extremely rare for any two companies to have the same strategies to achieve the desired promotional objectives. Hence every new appointee should receive training on the company structure, information on products including perceived advantages over competition, customer characteristics and administrative systems to service the order/delivery/ payment cycle. The time allocated to these activities will have to be extended if the company is hiring trainees or relatively inexperienced sales personnel. These latter individuals should also be exposed to a structured training programme to develop the basic skills of personal selling.

The majority of salespersons spend a large amount of time away from the company offices. It is vital that they exhibit a high level of self motivation and responsibility in the execution of their tasks, because their extended travel results in significant periods when they are beyond direct supervision. In recognition of this problem, the company should endeavour to motivate the sales force through mechanisms such as adequate remuneration, frequent communication and clearly defined career development opportunities.

The first decision concerning remuneration is that of the average earning expectation of a salesperson. Most companies base this decision upon the average level prevailing in the industry. Earnings should be related to defined roles and the actual payment an apparent measure to the employees of their performance. Of obvious attraction to the company are earnings totally related to sales, which can be achieved through paying a commission based on a defined percentage of business volume. The drawback to this approach is that the salesperson concentrates effort on immediate orders and is unwilling to spend time in other areas such as collecting information, developing new leads and improving customer service levels. Furthermore, poor market conditions can cause sales to deteriorate and employees paid on a commission basis may begin to worry about personal financial security and reduced morale could impair performance. Because of these issues, one tends to find the commission-only system in the fish industry restricted to companies in the commodity areas of business. Here generating orders is the only role of the sales force and these individuals have a large number of potential customers all of whom are willing to negotiate by telephone or telex.

The complete opposite to a commission payment system is to place the salesperson on a fixed salary without offering any incentive to motivate performance. The advantages of this method are the simplicity of administration and the total management control over the direction of salesforce effort. It appears to be a functional system if a salesperson has minimal involvement in order generation and is responsible for other, more administrative, tasks within the sales operation. Also, in some industries where there is a high technology component in the seller-buyer relationship, possibly combined with a lengthy negotiation between initial contact and final sale, it may be the most appropriate form of compensation. However, neither of these two situations occurs very often in the fish industry. The exception could possibly be companies in which the personnel are involved in managing sales training programmes or specialized tasks such as developing merchandising programmes for key customers.

Most companies find a mixed salary and performance bonus to be the most effective method of remuneration. A base salary is paid enabling management to specify sales force tasks and providing a base of financial security for employees. To this is added some form of bonus, with payment reflecting actual performance compared to specified sales volume or other objectives. The variety of potential tasks is almost infinite and hence the type of bonus scheme is extremely varied.

Material gain is rarely the sole source of motivation for individuals in business. Most people have career objectives related to their desire to advance to a position of greater responsibility and status. Sales staff are no exception, and the company should ensure that they understand these opportunities for advancement and that training programmes are provided to ensure that, when promoted, a person can execute the responsibilities of the new position.

Just as important in motivating performance is the factor of communication. Sales personnel are seldom in contact with individuals in their own and other departments. It is therefore vital that effective channels of communication are created which are designed to enable the sales force to carry out their responsibilities in the full assurance of efficient support from other members of the organization.

If a salesperson is frequently diverted from normal activities because of a customer complaint about delays in delivery due to errors in the traffic department, this individual cannot be expected to fulfil the sales objectives he has been assigned. Another component of communication which must receive attention is the public recognition of good performance. For salespersons who are rarely in the company offices, there is no greater reward than upon arrival at a meeting to be personally congratulated on a specific recent event associated with their actions in the field.

Sales control systems

The importance of the task and the long periods spent without direct supervision, cause control systems covering performance to be mandatory. Standard methods involve examining an individual relative to average activity, and the generation of both these data components will require each salesperson to complete reports on a regular basis. By nature, sales personnel regard report preparation as a tedious task which diverts them from their primary role. Hence it must be made apparent to them that the reports are being used not as a record of achievement but to identify areas of weakness which, through training and guidance, can be remedied. Any report of this type should be designed to be completed with ease and therefore should not contain requests for information which will not be utilized in the assessment process. The data collected will usually cover weekly or monthly quantitative information on:

(*a*) distance covered and/or time spent travelling to cover the territory
(*b*) number of appointments with existing customers
(*c*) number of orders taken from existing customers
(*d*) number of new accounts called upon
(*e*) number of first time orders from new accounts.

These reports can then be used in an analysis of activity relative to overall performance by calculating efficiencies such as

$$\text{Travel efficiency} = \frac{\text{Individual distance covered}}{\text{Average distance travelled}}$$

$$\text{Call efficiency} = \frac{\text{Individual number of existing customer calls}}{\text{Average existing customer calls}}$$

$$\text{Order efficiency} = \frac{\text{Individual number of orders}}{\text{Average orders per salesperson}}$$

$$\text{New account efficiency} = \frac{\text{Individual new account calls}}{\text{Average number of new account calls}}$$

$$\text{New account conversion efficiency} = \frac{\text{Number of new account orders}}{\text{Average new account orders}}$$

Examination of a salesperson employed by a catfish farm/processing co-operative may reveal that an individual has a high ratio for travel and call efficiency but a low rating for order efficiency. This would indicate possible weakness in persuading customers to buy and a possible solution would be to assign a senior salesperson to travel with the individual to provide training on 'closing the sale'.

Merely achieving a high order frequency is of no benefit if the orders are small or consist of a product line sales mix of low profitability. Evaluation of an individual's performance can be achieved from the accounts receivable records by assessing ratios such as

$$\text{Order size efficiency} = \frac{\text{Individual's average order size}}{\text{Company average order size}}$$

$$\text{Profit efficiency} = \text{Individual's average profit per order}$$

Identification by the management of the catfish co-operative of an individual with low scores for these ratios, would again indicate that remedial training should be undertaken to improve performance. For it is more effective to generate a few large profitable orders than for a salesperson to expend excessive efforts which only result in small orders from a scattered group of relatively unimportant customers.

The above 'salesperson to total efforts' comparisons are a vital mechanism for identifying a potential skills weakness. It is also necessary to carry out assessments on a time basis by comparing an individual's current record against past activity. This can be achieved by ratio analysis of such factors as current call frequency, call to order conversion rate, order size, and expenses versus prior year results. It should be recognized, however, that a poor score for any ratio could be related to an overall market decline in the sales territory beyond the control of the salesperson rather than a poor performance by the individual concerned.

Industry sales force constraints

The industry characteristic of being composed of numerous small companies which lack the individual resources to support advertising activity is also a constraint to supporting sales forces capable of providing adequate customer coverage. Rising energy costs have further compounded the problem because of the heavy travel component associated with the sales function.

In many countries this has resulted in companies moving towards a system where standard customer calls are managed by agents (often referred to as brokers) and the only paid sales employees are regional managers responsible for directing a number of brokers in an assigned territory. Some of these brokers accept responsibility to the point of purchasing product and adding a mark-up prior to resale. Although this role reduces inventory costs for the company, the relationship with the broker is changed because the latter is acting as an intermediary over whom the company has less control. It is therefore preferable to pay only an agreed commission on generated sales because this will cause the agent to concentrate on generating the maximum possible orders for the company and not, as in the alternative situation, expend effort to maximize profits on purchased inventory.

The selection, evaluation and control of suitable agents should receive as much attention as if these individuals were company employees. Therefore procedures described earlier in the management of a sales force are quite applicable to the operation of an agency network. In the resolution of performance problems, the company should recognize that the agency has a divided loyalty. Usually when acting to fulfil the best interests of the company, the agent will benefit from maximum possible earnings from commissions, but situations may develop in which the reduction of sales to certain customers would be in the best interests of the company. In these instances, following company direction could damage the profitability of the agent's operation. Thus in applying actions to improve agency performance, the company should recognize the need to generate solutions which are optimal for both parties in the company-agency relationship.

It is often assumed that the agent can never totally compensate for the advantages of control present within a company sales force system and this will be reflected in a higher sales volume from the latter structure. Thus a number of larger fish companies use agencies in poor performance

114

or low potential markets, but have a company sales force managing the more important markets. One can justify this logic in terms of cost because at a very high level of total sales, the payment of an agency commission will exceed the costs of a company sales force. In many cases, however, where a company has switched from an agency to a sales force, sales have declined. The initial reaction is often caused by customers switching to alternative suppliers in sympathy with the agent's loss of revenue, but the failure by companies to recover lost business over the long term cannot be attributed to the sympathy factor. The explanation is that the agent will usually have a more effective relationship with customers than can be created by most company sales personnel. The agent is perceived as a more permanent component of the distribution channel, whereas sales personnel are seen as potentially transient, switching areas and even employers depending upon circumstances. Consequently, the agent/customer relationship involves greater trust and clarity of communication which will be reflected in a size and stability of business rarely matched by a company salesperson. The reality of this situation is gradually becoming recognized, and it is likely that in the future greater attention will be given to the management of company-agency systems accompanied by de-emphasis of large company sales forces interfacing directly with customers. However, there is always the personal obstacle that many sales directors believe their status in a company is greater when they are responsible for a large number of sales personnel as opposed to a few regional managers supervising an agency network.

11
Planning

A frequently mentioned theme of earlier sections is that marketing is a functional role concerned with the process of managing change. The achievement of effective and consistent results is only possible by placing emphasis on planning. Formalizing the planning process encourages systematic forward thinking, leads to better co-ordination of effort, creates standards against which performance can be measured and equips management to handle environmental change. The combined effects of these activities is reflected in statements of corporate objectives and strategies by which to guide overall performance.

The mechanisms of the planning process vary widely between companies, but can be divided into three main types:

(a) 'Bottom-up objective and planning' — where lower management develop a full proposal for approval by senior staff.
(b) 'Downward planning' — in which only senior management are involved in the process and their final proposal is transmitted to lower management for execution.
(c) 'Objectives downwards/plan upwards' — which is a combination of the two above approaches, with senior management formulating specific objectives that are given to lower management for response with a formal statement of strategies and policies through which to achieve the specified corporate objectives.

The planning process is often divided into two phases, long-term and annual plans. The former involves the development of basic direction to guide the company efforts over the longer term, typically defined as a three to five year period. From this process can evolve the framework around which the annual plan, designed to achieve specific goals and define policies, can be centred.

The logical starting point for the long-term plan is the diagnosis of the company's current market position and the factors which have contributed to this situation. Much of this data will be contained in existing information available within the marketing group. Facts concerned with operating costs and capital equipment investment requirements may be provided by the accounting group. From this a prognosis on the future can be developed, namely, 'Can the company sustain current sales and/or profit trends in the current market or will internal and/or external trends require a revision in strategy?' A useful starting point in this form of assessment is to forecast both industry and company sales. The sales forecast, when linked with an assessment of the product costs and marketing expenditure required, can provide a profile on future profit expectations. Ultimately the objective is to generate profits sufficient to sustain or exceed a specified return on investment (ROI). The asset base necessary to support the sales forecast with, if necessary, additional capital investment for plant renewal or volume expansion pruposes, will also be examined. Should this total review reveal future problems over performance, the company will be forced to examine alternative strategy paths. These may involve merely a move towards increased market penetration for existing products or at the other extreme demand a major change in direction, possibly to the point of diversification into areas totally outside the company's current marketing operation.

To illustrate the concepts of long-term diagnosis, prognosis and marketing strategy revision, let it be assumed that a medium sized trout farm operation has developed a forecast which indicates:

(a) Some opportunity for further industry expansion in the retail market for whole trout, possibly +15%.

(b) Expected output expansion by competition will place pressure on prices.

(c) To sustain current market share would involve an increase in marketing activity with a consequent reduction in profitability and ROI.

These conclusions are based on forecasts developed by the marketing group and cause management to conclude that a sustained ROI of 25% (let alone attainment of a previously defined long-term objective ROI of 30%) will require review of other strategy paths. The lowest risk proposition is to consider alternatives within the aquaculture industry because this is the current area of corporate expertise. Possible routes of new technology, new species and/or new markets arc described in *Fig 11·1*.

117

| | | TECHNOLOGY DIMENSION | |
		Related Technology	Unrelated Technology
D			
E	Existing	New culture species	New culture species
M	Segment	utilising similar	utilising different
A		production technology	production technology
N			
D	Vertically	Manufacture feed	Further processing of
	Adjacent	materials	culture species
S	Segment	Manufacture of capital	harvest
E		equipment	Market branded goods
G		Disease control systems	Open end-user outlets
M			
E	Similar or	Agricultural crops	Sales agent for
N	Related	Terrestrial live-stock	commodity fish and
T	Segments	production	seafood products
			Trawler and
D			processing operation
I			
M	New	Pollution control	ANY INVESTMENT
E	Segments	Natural resource	OPPORTUNITY
N		management	
S		Agricultural disease	
I		control technology	
O			
N			

Fig 11·1   Alternative strategy paths for the trout operation

The annual plan

If the long-term plan is sufficiently precise, the annual plan can focus in detail on the strategy and policies required to deliver the financial objective for the next twelve months. Without the former process, an annual plan can often deteriorate into a set of ill-chosen solutions to short-term issues which reflect attempts to remedy last year's problems and predict possible new problems to confront the company during the next few months. This does not mean, however, that the well-organized planning

118

process will not include a review of prior performance. Typically a starting point in any annual plan is an evaluation of actual performance versus forecast for sales, cost of goods, gross profit, marketing expenditures and net profits. If variance is identified, negative or positive, it is beneficial to develop an explanation of these in order to avoid recurrence of the same errors in the future.

If there is a general acceptance of the specified financial goals for the coming year, the attention of the planning process can be focused on quantitative validation of financial targets and defining an optimal marketing mix. In practice, the plan is not evolved as a linear action moving along a sequential track, because frequently an obstacle at one point may require review of earlier conclusions. For example the advertising programme may assume achievement of 80% product distribution at store level, whereas analysis of the personal selling programme might indicate that this is excessively optimistic. If a lower distribution figure is accepted the validity of the advertising programme would then have to be reviewed.

Planning is therefore a somewhat circular process and the product component of the marketing mix is the most practical point of entry. Assuming a reasonably constant industry sales trend, the product decisions can be focused on the company's expected share of business. This requires assessment of both actual product performance, *eg* Are any recently introduced products exhibiting signs of failing to satisfy the benefit claim for which they were originally formulated? Can products in the maturity phase be expected to retain a sufficiently unique positioning, or is reformulation/re launch necessary? Which items in the declining phase should be scheduled for orderly discontinuation?, and assessment of potential actions by competition in relation to possible impact on any area of the product line, *eg* a product previously in a unique position with promotional support emphasizing unduplicated benefits now faces a competitive entry which could require a redirection of promotional effort towards communicating a product superiority claim over competition. The product line examination can be more effective if the company has an on-going programme of research on market performance and consumer behaviour. When these data are on file, one has only to extrapolate substantiated trends to reach an accurate prediction of share attainment. Without this supportive material, the company's decision process is based on minimal knowledge and high personal judgement with the consequent risk of incorrect performance estimates.

119

Most companies utilize a 'follow the market' pricing system and/or are sufficiently marketing orientated to consider price as only one component of the marketing mix, so that the issue of price may not require detailed deliberations. Nevertheless careful assessment of probable price levels is required because this factor will influence sales volume. The marketing group tends to emphasize sales revenue, but other departments will need to derive unit sales forecasts as the basis for decisions on raw material purchase, plant throughput and inventory management. Therefore a relatively high degree of accuracy is expected of the marketing group in their statement of expected prices. If, however, the marketing group encounter a profit margin problem for specific products, a more detailed price review will be instigated to determine whether opportunity exists to reduce cost of goods and/or adjust price to solve the profit problem.

If the marketing group has established a quantified relationship between sales and promotional experience, it is a relatively simple process to prepare an overall promotional budget. All too frequently marketing groups do not have a sales/expenditure response model and tend to set budgets similar to the previous year unless future financial goals necessitate a revision in promotional activity. When net profit is trending downwards, promotional activity is often reduced in the annual plan. This could merely compound the problem by dampening future sales, thereby placing further pressure on profits. If net profits are in excess of objective, the promotional expenditure may be increased, and in certain situations these incremental funds will not generate an equivalent return in terms of further profit growth.

Thus in the absence of a sales/expenditure response model, it is worthwhile to execute a subjective examination similar to the type described for establishing advertising budgets. The marketing group should review the effect on market share if expenditure remains unchanged, and also if increased or decreased by various amounts, eg ± 50%, ± 20% and ± 10% versus current levels. These forecasts can then be restated in relation to revenue, expenditure and profit to see if an optimal promotional programme should be introduced.

Having defined the promotional budget, the next stage is to decide on the allocation of funds relative to advertising, personal selling and below-line activity. In many companies below-line programmes are considered as an alternative only to advertising and hence the primary budgetary split is between advertising plus below-line programmes and personal selling. The industrial market sector will tend to allocate the majority of funds to personal selling whereas in the consumer sector this will tend to be a minority expenditure component. In either situation, selling expense will usually contain an element of commission either paid to the sales force or the agents, and this expense is directly related to sales revenue. This variable component should be estimated and deducted from the promotional budget prior to allocation to any other programmes.

Once the advertising budget has been established it is usually forwarded to the advertising agency, which will be more qualified to develop the advertising plan most appropriate to the overall promotional objectives and funds available. Below-line programmes have only short-term influences but again will reflect overall objectives. For example, if increased trial is a primary goal, the funds will be directed to actions such as free samples and high value coupons to subsidize initial customer usage. Where increased repeat usage is desired, this will be reflected in the below-line programmes such as trade allowances per case to stimulate intermediaries to organize merchandising activities.

Distribution is a long-term issue to which major attention is rarely given in the annual plan, except to review the current level of distribution achieved in each channel. From this definition, priorities may be developed for use in briefing the sales force on areas which have not attained an optimum level and will require attention in the coming months. There is little point in mounting advertising campaigns which utilize mass communication if the customer cannot locate the product at point of purchase. Hence many companies have previously established targets, *eg* 70% distribution in retail stores, below which advertising campaigns will not be executed until the distribution has reached the defined minimum level.

The annual plan, as well as being the framework for the overall marketing operation during the next 12 months, is also the basis for defining effort within each of the company markets across the country. Consequently another component of the plan will be to define sales targets and promotional plans for each sales territory. The balancing of the goals and programmes at market level will depend upon whether the marketing group wish to emphasize a sustained sales level in existing stronger territories or to generate growth through exploiting potential volume opportunities available in the less well developed territories. Whichever is the chosen route, the decision is made more effective if the company has developed understanding of the relationship between promotional expenditure and sales at territory level. The company can then evaluate the optimum allocation of funds to generate the best possible return. For example, in certain territories, any additions to current expenditure may have little impact on generating incremental sales. Other territories may have been allocated spending levels which are so small they would have no measurable impact on performance. A third group of territories may only require an injection of additional funds to generate major sales growth. Under these circumstances funds in the first two territory groups would be more efficiently utilized if transferred to the third.

Given the need for the company to be able to react to change, the plan should also contain references both to total plan alternatives and the testing of new propositions.

The 'total alternative' is usually a contingency plan for implementation should actual events not occur as forecast. This could in theory result in an infinite number of possibilities each demanding a solution, but for most companies the only practical approach is to examine only two plans. One examines the implications of a major upswing in market demand which generates incremental profits for reinvestment in the marketing operation. The second deals with a dramatic decline in market demand and a consequent proportionate cutback in corporate marketing programmes.

The testing of new propositions should encompass market/customer evaluation of alternative promotional plans and concepts through small scale test of variations in spending levels or promotional mix allocations. New promotional concepts could include a revised advertising campaign emphasizing different product benefits and below-line programmes, *eg* a temporary value incentive linking the products with another company as a joint in-store merchandising event. The tests of overall spending or promotional mix allocation is designed to provide an ongoing validation of the company sales/expenditure response curves and optimal advertising/personal selling allocation decisions. These activities, when managed effectively, will mean that the company is rarely unprepared for sudden changes in environmental conditions. Should a variation in forecast occur, the company can introduce either a new total plan or a revision to a subcomponent of the marketing mix, both of which have already been evaluated in relation to expected sales, expenditure and net profit.

Simplified example plan

Long-term strategy
Some aspects of the annual plan development process can be exemplified by a hypothetical fish and seafood processor marketing a range of

122

branded goods in both the retail and food service sector. Two years earlier, senior management was changed because of shareholders' dissent over the poor return on investment (ROI) performance of only 12%. The new management identified two problems: first the low profit margin on unprocessed products; and secondly the seasonal nature of catch (even though the company did not operate trawlers), linked to a need to sustain a reasonable customer service level on a year round basis, which demanded a raw material inventory investment equal to five months' sales. When this is added to the finished goods inventory equal to two months' sales, then almost 70% of the company's assets would be tied up in servicing inventory needs.

The revised long-term objective was to attain a corporate ROI of at least 20%. This would be achieved by a strategy of phasing out poor margin unprocessed products and replacing the sales with higher margin processed items. These latter products, because the finished goods only contained 50% fish or seafood flesh, would also require a lower total inventory investment.

Annual performance review

The combined total industry sales of branded frozen goods and food service of £200 million (in factory level sales) is made up of the following:

Industry Annual Sales
£('000,000)

Category	Retail (% *Importance*)		Food Service (% *Importance*)		Total (% *Importance*)	
Raw fillets	30	(30·0)	50	(50·0)	80	(40·0)
Raw seafoods	40	(40·0)	20	(20·0)	60	(30·0)
*Processed fish	15	(15·0)	10	(10·0)	25	(12·5)
Processed seafood	15	(15·0)	20	(20·0)	35	(17·5)
	100	(100·0)	100	(100·0)	200	(100·0)

*In all the following Tables the products are covered in either batter or breadcrumbs or forming the base for entrée type items

No change is expected in total industry volume during the forthcoming year. At the end of this year, the company expected to have a 10% share of total industry sales, although the sales mix relative to the industry, is (*a*) biased more towards the retail sector and (*b*) reflects efforts to reduce sales of raw products.

123

Company Annual Sales
£('000,000)

Category	Retail (% Importance)		Food Service (% Importance)		Total (% Importance)	
Raw fillets	5	(33·3)	0	(0·0)	5	(25·0)
Raw seafoods	5	(33·3)	2	(40·0)	7	(35·0)
Processed fish	2	(13·4)	2	(40·0)	4	(20·0)
Processed seafood	3	(20·0)	1	(20·0)	4	(20·0)
Total	15 (100·0)		5 (100·0)		20 (100·0)	

Share of industry sales 15·0% – 5·0% – 10·0% –

Sales are expected to close on budget with expenditure and net profits forecast to be at the following levels:

Company Profit and Loss £('000) Statement

		(% of Sales)
Total sales	20,000	(100·0)
Gross Profit		
Raw fillets	600	(12·0)*
Raw seafoods	1,085	(75·5)*
Processed fish	985	(24·6)*
Processed seafood	700	(17·5)*
Total gross profit	3,370	(16·9)
Expenditure		
Sales expense **	600	(3·0)
Advertising	300	(1·5)
Below-line programmes	200	(1·0)
Admin. general expense	600	(3·0)
Total expense	1,700	(8·5)
Net profit	1,670	
Average asset base	8,932	
Return on investment	18·7%	

*Gross margin expressed as % of individual category sales
**Expenditure directly related to sales and therefore considered expense element

Annual objective, strategy and forecast
The objective for next year is to attain the long term target of a 20% ROI.

This will be delivered through continued emphasis on processed goods with concurrent reduction in sales for raw products. Given the importance of retail volume, the company will endeavour to provide advertising support for the more unique processed goods with a secondary objective of building consumer awareness for the brand name across the entire retail product line. Competitive actions are expected to remain relatively unchanged. This situation, senior management's refusal to consider investment spending on marketing programmes which would mean a reduced net profit over the short-term, will cause overall promotional expenditure to remain at current year levels.

This strategy is forecast to achieve the following financial results during the next 12 months:

	£('000)	(% Retail)	(% Food service)
Category Sales			
Raw fillets	4,000	(100·0)	(0·0)
Raw seafoods	7,000	(73·3)	(26·7)
Processed fish	4,500	(55·5)	(44·5)
Processed seafoods	4,500	(75·0)	(25·0)
Total sales	20,000	(75·0)	(25·0)
Gross Profit			
Raw fillets	480	(100·0)	(0·0)
Raw seafoods	1,085	(87·3)	(12·7)
Processed fish	1,107	(75·7)	(24·3)
Processed seafoods	828	(75·8)	(24·2)
Total profit	3,500	(86·3)	(13·7)
Expenditure			
Sales expense	600	(75·0)	(25·0)
Advertising	350	(100·0)	(0·0)
Below-line programmes	150	(66·6)	(33·3)
Administration/ general expenses	600	(70·0)	(30·0)
Total expense	1,700	(77·6)	(22·4)
Net profit	1,800	(94·5)	(5·5)
Average asset base	8,580*	N/A	N/A
ROI	21·0%	—	—

*Reflects reduction in raw product sales mix

Promotional plan

The need to build the retail business through the major stores and large wholesaler/smaller retail unit distribution system requires as much emphasis as possible on advertising. Nevertheless, 55% of total funds will be expended on personal selling which reflects the fact that the company employs a broker force paid by commission, and that this expense cannot be reduced without damaging sales revenue.

Advertising will focus on the newer processed fish products introduced last year, with approximately 70% of expenditure on television advertising in selected market areas and the balance on newspaper advertising linked to major supermarket merchandising events. Although the marketing group have yet to establish a precise sales/advertising expenditure response course, on a judgemental basis it is felt that advertising across all sales regions would dilute message impact to the point of generating only minimal return. Advertising will therefore be restricted to the three sales regions where the company has a strong sales base and retail distribution is in excess of 70%. The lesser importance of food service over the near-term means that no advertising will be executed in this sector of the business.

The marketing group, in recognition of the need for more information on sales/expenditure response courses, intend to allocate £7,000 of advertising and £3,000 of the below-line budget to test the impact of a 50% spending upweight in a sub-area of one sales territory. The results will be measured by evaluation against normal promotional activity and another sub-area of zero spending.

Although it is understood by the marketing group that below-line programmes are only of short-term benefit it is felt that this mechanism is required in order to generate merchandising support by intermediaries while also encouraging both trial and repeat purchase by consumers. Of the available funds, £20,000 will be used on trade allowances in the food service sector to reduce case price on key items during specified promotional periods throughout the year, and the balance will be directed into the retail sector split into proportions of 70:30 between trade allowances and consumer coupons 'good on next purchase' of processed fish or seafoods. The coupons will be delivered through the planned newspaper advertising. Trade allowance activity will be restricted to processed goods and the higher margin raw seafoods. Given the intention to reduce sales volume for raw fillets, no promotional funds will be allocated to this category.

Pricing and distribution
Based on information from the accounting and procurement departments, it is expected that costs will rise in line with inflation. The company is a price follower and it is assumed that there is little risk of adverse competitive price reactions as the company moves overall prices upwards during the coming year. The price increases can be expected to reduce the company's total sales volume. However, as the major components of fixed cost are absorbed by application to processed goods during manufacture, and these categories are positioned for incremental growth, the decline in overall volume is not expected to produce any burden absorption problems in the next twelve months.

No major alterations are planned in the company distribution channels. It is thought, however, that the sale of food service products direct to major restaurant chains under the intermediary brand name is more cost-effective than marketing own brand products through small, widely dispersed wholesalers. It is proposed, therefore, to examine ways to increase sales of intermediary brand processed goods during the coming year.

Territory allocations
The bias towards support in the three strongest sales territories is expected to be reflected by some incremental growth in these areas. The weaker sales territories can be expected to exhibit further decline in response to minimal promotional support. This conclusion is reflected in the regional sales forecasts.

| | £('000) | | | |
	This Year	*(% Total Sales)*	Next Year	*(% Total Sales)*
Region 1	2,200	(11·0)	1,800	(9·0)
Region 2	1,000	(5·0)	900	(4·5)
Region 3	6,000	(30·0)	6,300	(31·5)
Region 4	4,000	(20·0)	4,200	(21·0)
Region 5	5,000	(25·0)	5,200	(26·0)
Region 6	1,800	(9·0)	1,600	(8·0)
Total	20,000	(100·0)	20,000	(100·0)

The key objectives for the sales force at regional level are to sustain achievement of a 70% distribution level in the regions to receive advertising, and also to ensure that the expected sales declines in the remaining regions do not cause trade discontent. For although it is planned to reduce promotional emphasis in these areas in the near-term, at a later date the company expects to expand processed goods sales volume in these markets. This will not succeed if the company image has been previously damaged amongst the intermediaries during the phased reduction in unprocessed fish and seafood sales.

New products and alternative plans
The marketing group believes that efforts should be directed towards a new range of improved coating performance processed seafoods and a range of up-market seafood entrées. This reasoning is based upon the fact that competition is much stronger in the processed fish category, and therefore new processed seafood items represent an easier market entry proposition as well as helping to create a more distinctive image in the market. The preliminary phase of the programme, concept development, prototype design and consumer evaluation, estimated at £50,000, is to be funded from the general marketing expense budget. It is hoped that this work could result in the introduction of the first new items within 12—18 months.

The practice of alternative plans is not a concept which has been practised in the past. Senior management now recognize the benefits of this practice and have requested the preparation of a skeleton framework for an alternative plan. Further it is intended to demand more detailed alternative plans in future years. The marketing group was therefore asked by senior management to provide a short statement in response to two planning scenarios: (*a*) a reduction in gross profit of 25%, and (*b*) an increase in marketing funds of £40,000.

The reduction in gross profit of 25% means a loss of £875,000 in absolute profits. It is assumed by the marketing group that it would be caused by a proportionate reduction in sales volume, which would be reflected by a reduction in variable selling expense of £300,000. They also conclude that the only further savings in expense which could be made would be to cancel all advertising. Of the £350,000 savings this would achieve, £50,000 would have to be directed into added below-line activity to reduce intermediary discontent about the lack of advertising support and to sustain consumer interest over the short term.

The consequent profit statement for this alternative plan is forecast as follows:

	£('000)
Sales	15,000
Gross profit	2,625
Expenditure	1,100
Net profit	1,525
Average asset base	7,722*
ROI	19.7%

*This figure assumes that the reduced sales level will be reflected by a lower inventory investment

The low spending level in the alternative plan could not be sustained for more than a short period without placing the company franchise in the market at risk. The marketing group should put a footnote to this effect in their proposals in order to prevent these expense savings from appealing to senior management more than the full scale promotional plans previously presented for approval.

The alternative scenario of an increase in expenditure of £400,000 is assumed to be a permanent, not one year only, increment by the marketing group. They recommend that the funds be directed towards progressing the move from unprocessed to processed goods. Areas of investment should be partially immediate, with £200,000 on advertising processed fish items in Regions 1 and 6. Region 2 is considered too undeveloped for increased promotional activity to be of benefit. £100,000 should be invested in accelerating the introduction of the new processed seafood products. Another area of opportunity lies in the food service sector and one current weakness is insufficient staff involved in key account selling. £50,000 should be allocated to hiring more sales staff and a further £50,000 on research and development for a range of more unique, higher profit margin goods for the food service sector. Current products lack sufficient differentiation from competition and thus price competition, with the consequent reduction in gross margin, is a major element of the present product positioning.

The increased advertising expenditure is the only component which would have an immediate impact, generating an estimated £1 million in additional sales. After deducting 6% of sales for variable selling expense this will represent an incremental net profit of £190,000 (or almost break even). The other proposals would not have any immediate impact, but over a three year period are forecast by the marketing group to result in an increase in processed goods' sales of over £5 million at a 25% gross profit margin.

12
Control systems

The introduction of a structured planning process in a company is of little benefit if it is not accompanied by an evaluation of performance, followed where necessary by remedial action. As modern commercial organizations are becoming increasingly complex with employees assigned to specialist tasks and the availability of information rising at an exponential rate, it has become mandatory to manage evaluation and reaction formally through the use of a control system.

The four basic components of a control system are: establishing standards, evaluating performance, analysing variance and instigating corrective actions. The other vital component without which the control system cannot function effectively, is feed-back to assess the impact of corrective actions on future performance. This feed-back loop may also be used on a less frequent basis to evaluate whether failures of performance reflect incorrect assumptions during the original planning process, rather than poor execution.

The high level of expenditure by marketing departments relative to other groups within most companies should mean that the former have the most sophisticated control system in operation. However marketing personnel appear to resist the introduction of even the most minimal of control structures. Their justification for this position is that the complexity of the interaction between the variables of the marketing mix cause control systems to be of little benefit. During the last twenty years, as marketing departments have become responsible for the majority of total corporate expenditure, senior management have started to insist on the more effective control mechanisms within this area of company operations.

Most control systems are based upon accounting techniques developed by financial managers, which involve periodic reviews of performance trends or cost effectiveness. Although this is an effective approach, there is a growing awareness that marketing is a continuous process which requires more sophisticated methods. The development of such control systems has been made feasible by the availability of computer technology and the associated improvement in mathematical analysis techniques for business situations. The resulting approach, known as Marketing Information Systems (MIS), is now being introduced into the larger, marketing orientated companies, but as yet has received very limited acceptance in the fish industry. Some aquaculture groups have introduced computer decision models to manage their production processes more effectively and it is likely that these companies will also be the first to adopt the concept of MIS.

The components of the MIS are described in *Fig 12·1*. The system combines the collection of external and internal information, identification of areas requiring specific data to be obtained through market research, all these sources becoming the primary input for computer models to predict the outcome of alternative decisions. Having then selected an optimal decision, the most appropriate remedial solution can be executed by the marketing group.

In some companies, the modelling process has evolved to the point where marketing groups can evaluate changes in variables, *eg* a revision in advertising expenditure relative to sales, without the need for any validation through test market activity. Other models can assume the viability of new product concepts without market placement to evaluate sales potential. Such techniques are now at the point where first year sales volume can be predicted with an accuracy of less than ± 10%. It is obvious that a company with the capability in the fish industry has significant management superiority over competitors who have not introduced MIS into their operation.

For the majority, however, MIS is a concept of the future. In the meantime the traditional approach to control through performance evaluation or efficiency analysis will be employed, and for a more in-depth examination of the total operation a 'marketing audit'.

Performance evaluation

The annual plan provides a precise definition of the quantified objectives for sales, expenditure and profit. These data should then provide a basis for evaluating the validity of achieved performance during the year.

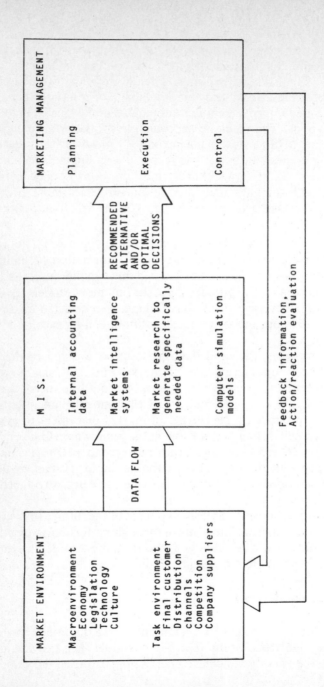

Fig 12.1   The marketing information system and marketing operation

Sales analysis

The variance between planned and actual sales can be due to a difference in the unit sales volume and/or unit price. The relative importance of these two variables can be assessed through variance analysis using the formulae:

Variance due to price =
(Forecast price — actual price) × actual volume

Variance due to volume =
Forecast price × (Forecast volume — actual volume)

Actual sales performance is not only a reflection of company marketing activity but also a reflection of the environmental conditions within the market. Sales should also be assessed, therefore, in terms of the company share of total industry sales. A company may be achieving a forecast +10% increase which could cause some complacency in the marketing group. However, if this is accompanied by a decline in the company share of total industry sales, it can be assumed that total industry volume is increasing but that company growth is less than industry growth rate. Where company and industry growth rates are similar, it can be expected that company share will remain unchanged. Alternatively the company sales may be 10% below target, but share of industry sales up by 20%. This would indicate an overall decline in total market size with the company marketing programmes partially counteracting the influence of the environment on company performance.

Where the company markets a range of products, the control system should include an analysis of sales versus forecast at the individual item level. For example the hypothetical fish company plan described in the earlier section may have a month 1 forecast of £1·5 million and actual sales are £1·47 million. Upon examination of each item, it is apparent that the company is failing in the key area of processed products and generating excessive sales for unprocessed items.

	Month 1 £('000)		
	Forecast	Actual	% Variance
Unprocessed fillets	420	500	+19·1%
Unprocessed seafoods	380	435	+14·5%
Processed fish	500	375	−25·0%
Processed seafoods	200	160	−20·0%
Total	1,500	1,470	−2·0%

Profit analysis
The net profit is influenced by total sales, cost of goods and expenditure as described by the formula

Net profit = Sales revenue − (cost of goods + marketing expenditure)

If the sales analysis indicates no problem in this area, then a below forecast net profit can be due to a fault in the cost of goods and/or excessive marketing expenditure. The cost of goods issue requires an analysis to see if this reflects an incorrect product mix, *eg* the earlier example of the fish company where the below forecast sales of higher margin processed goods will reduce overall profit, or a cost of product in excess of the production/procurement department's expectation. In the fish industry, the latter is not uncommon because predictions of fish flesh costs assume a certain level of market demand and supply. As prediction of actual world landings is almost impossible, the industry frequently encounters the problem that shortages develop, place pressure on landed prices and, in turn, profitability per unit of sale is reduced. In this situation the marketing group will have to review the possibility for price increases to recover margin. If a price increase would reduce customer demand and hence overall sales revenue, the only alternative is to examine whether marketing expenditure can be reduced.

Expenditure analysis
A logical start point in expenditure analysis would seem to be a review of total expenditure versus budget, but in practice this approach will provide minimal information. The important factor is the relationship between expenditure and sales, which can be examined through an analysis of expenditure/sales ratio. This can be illustrated by an example of the quarterly regional performance of the example fish company. The forecast for the second quarter is:

Region	*Sales*	*Advertising*	*(% Adv/Sales)*	*Below-line*	*(% BL/Sales)*
			£('000)		
1	450	2·0	(0·5)	3·2	(0·7)
2	225	0	(0·0)	2·0	(0·8)
3	1,575	32·8	(2·1)	11·5	(0·7)
4	1,050	32·0	(2·1)	7·5	(0·7)
5	1,300	26·9	(2·1)	9·3	(0·7)
6	400	2·0	(0·5)	3·0	(0·8)

The actual performance for the second quarter and resultant expenditure/ sales ratios was:

	Sales	£('000) Advertising	(% Adv/Sales)	Below-line	(% BL/Sales)
Region					
1	490	2·0	(0·4)	3·4	(0·7)
2	200	0	(0·0)	2·0	(1·0)
3	1,165	32·1	(2·8)	11·8	(1·0)
4	1,650	21·8	(1·3)	7·6	(0·5)
5	1,380	26·6	(1·9)	9·4	(0·7)
6	340	2·1	(0·6)	3·8	(1·1)

By only evaluating actual expenditure versus budget, all regions show little variance from the forecast. For example, advertising by region actual versus budget: Region 1 0·0% variance; Region 2 0·0% variance; Region 3 2·2% over budget; Region 4 2·0% below budget; Region 5 1·1% below budget; Region 6 0·0% variance. However, when comparing expenditure/ sales ratios, it is clear that in certain regions the poor sales are causing an increase in the ratio, eg Region 3, the A/S ratio actual is 2·8% versus the budgeted 2·1% of sales, whereas in other areas, eg Region 4, an A/S ratio of 1·3% versus the budgeted rate of 2·1%, higher than forecast sales are reducing the expenditure/sales ratio. The choices available are to review mechanisms to improve sales performance in the poorer areas or reduce expenditure and redirect funds into areas where sales are exceeding budget.

Efficiency analysis

The most dangerous assumption in marketing is that current functional methodology is at maximum effectiveness because these issues are carefully reviewed within the annual plan. Given the importance of ensuring the company is utilizing the most appropriate marketing mix, an on-going control system should exist to examine each aspect of the 'four Ps'. Most of the analysis issues were handled earlier during discussion of the marketing function, eg risk reduction in new products; profit optimization during the product life cycle; sales/expenditure response curves and systems to measure sales force effectiveness, and every effort should be made to include such matters as a cornerstone of the management process.

The objective of these processes is to maximize financial performance through decisions concerned with optimal profit at minimal cost. In theory this should be evident from the Profit and Loss Statement for the company. Unfortunately this document is prepared for accountants as a description of overall activity and unless further analysis is carried out, the general nature of the figures will tend to obscure the very typical situation that certain areas of the marketing operations are less profitable than others. Unless this variance by area can be identified, the actions necessary to improve operating efficiency (whether in the form of a revision of activity, re-direction of resources or withdrawal from certain business segments) cannot be taken.

One standard approach is to break down the Profit and Loss Statement into separate sectors which describe areas of activity as individual profit centres. The form of classification will depend upon specific interest at the time of analysis, but some of the commoner approaches include an evaluation by product category, market types, geographic areas and customer types. A frequent problem is that existing accounting information will be inadequate, and in these instances the marketing group will have to take data on-hand and re-allocate them on the basis of constructed equations to reflect costs relative to expended effort.

To demonstrate this concept, assume our example company wishes to gain greater insight on the issue of marketing processed goods to food service intermediaries under the latter's brand name instead of distributing own label products through a large number of geographically dispersed wholesalers. Accounting information on the prior year performance and the business plan for the next year provide statements of overall sales revenue and gross margin by business type. The components of marketing and administration expense, however, will have to be restated to permit calculation of net profit by business type.

The accounting data from the prior year indicates that intermediary label product is sold at a lower price than own brand, with margins respectively being 9·5% and 14·2% for fish, 14·5% and 18·3% for seafood. Applying these margins to the sales projection for the next year and a target of 85% of sales under own label yields:

	£('000)		
	Intermediary label	Own label	Total
Sales processed fish	310	1,690	2,000
Sales seafoods	150	975	1,125
Total sales	460	2,665	3,125

	£('000)		
	Intermediary label	Own label	Total
Processed fish gross profit	29·5	240·5	270
Seafood gross profit	21·8	178·2	200
Total gross profit	51·3	418·7	470

The expense component that can be analysed immediately is the sales commission, because 3% is paid to brokers on company label but only 2% on intermediary brands. Also, of the below-line programmes, only £20,000 is spent on food service with all this in support of company brand processed goods.

The administration budget is not broken down beyond the following sub-grouping of:

	£('000)
Sales administration	100
Marketing	200
Customer service	30
Packaging development	50
New product development	100
Market records	50
Merchandising material/publicity	70
	600

Certain of these components can be excluded because they only cover the retail operation, namely market research, new product development and packaging. The balance have to be allocated on some reasonable basis.

Sales administration can be sub-divided using the sales report sheets to analyse proportion of time spent on various accounts. This yields the following:-

Retail	Food service
% hours 64	36
Food service wholesale	Intermediary label accounts
% hours 81	19

Allocated
costs = (£30,000 × 0·36 × 0·81)   (£30,000 × 0·36 × 0·19)
     = £8,748   £2,052

A similar analysis of hours yields the marketing administration allocated costs as

Food service wholesale   Intermediary label
£20,060   £13,940

Customer service is involved in solving customer delivery credit issues by interfacing with the respective departments in the company. The simplest allocation is to assume an average order size (in this case £8,000) and calculating number of orders by dividing total sales (20 million) by average order size = 2,500. Service cost per order is estimated by dividing total operating cost (£30,000) by total orders. This yields an average cost per order of £12. This figure can be used as the basis for estimating service costs as follows:-

		Food service wholesalers	Intermediary label
(1)	Sales	£2,665,000	£460,000
(2)	Average no of orders*	333·1	57·5
(3)	Service cost**	£3,997·2	£690

* = Sales ÷ £8,000;   ** = Orders × £12/order

Merchandising and publicity are allocated on the basis that only 15% of funds are expended on food service (0·15 × £70,000 = £10,500) and no merchandising or publicity is expended on intermediary label business.

Considering all the above, the following profit statement is generated for the two areas of business:-

£('000)

	Food service wholesalers		Intermediary own label business	
Sales		2,665		460
Gross profit		418·7		51·3
Expense sales commissions	80·0		9·2	
Below-line programmes	20·0		0·0	
Sales admin	8·8		2·1	
Marketing admin	20·1		13·9	
Customer service	4·0		0·7	
Merchandising/publicity	10·5		0·0	
Total expense		143·4		25·9
Net profit		275·3		25·4
Profit as % of sales		10·3%		5·5%

This analysis reveals that expense as a percentage of sales is similar for both areas (5·4% of sales for the company own label versus 5·6% for intermediary labels). The lower margin caused by lower prices in the latter segment causes net profit as a percentage of sales to be significantly better for the company label product. Hence the concept of concentrating on intermediary label may need further evaluation before significant business expansion is instigated.

The marketing audit

Performance evaluation and efficiency analysis tend to focus on near-term issues and hence do not provide an overall assessment of the total marketing operation from general strategy through to execution. A number of companies are now introducing the concept of a periodic marketing audit, executed by an outside consultant or through an internal review procedure, to gain insight on the totality of the marketing process.

The audit will usually focus on five main issues: the external environment, objective and strategic planning, the components of marketing mix, organization, and control. The execution of the audit will require collection of information on these issues followed by a formal review to assess which, if any, areas require revision to improve effectiveness.

The external environment should be examined relative to such issues as:

— What are the company markets?
— What major segments exist within each market?
— What is the current size of these markets and segments?
— Which represent areas of growth or decline?
— Who are the company's target customers?
— What is known about customer buying behaviour?
— What possible change in behaviour could occur in the future?
— Who are the competition?
— What are strengths and weaknesses of the competition relative to company performance?
— What change can be expected in competitive activity in the future?
— What are the relevant macro-environment variables of influence, *eg* technological, legal, cultural?
— Which of the macro-environmental variables can be expected to change in the future?

The objectives and strategies should be assessed in the context of their long and short-term components, as well as being reviewed in terms of their relevance to the future prospects for corporate developments.

Another important indicator is actual performance versus plan, to determine whether the marketing group is achieving specified financial goals assigned through adherence to the overall corporate philosophy established by senior management. It is not unusual to discover that to meet performance standards, the marketing group is stating one strategy to placate senior management but in fact is executing another, non-articulated strategy which they believe is more feasible in current economic circumstances. This dichotomy of opinion is of no use to either party and should be modified accordingly.

The components of the marketing mix or 'four Ps' will involve issues of:

— What are the current major products?
— Are current products correctly positioned?
— What are the implications of the life cycle curve on future revenue?
— Should any products be discontinued?
— What new product developments are being pursued?
— What is the success/failure rate for new products?
— How are pricing policies established?
— What revisions in pricing policy could improve overall profitability and/or profit margins?
— What are the current distribution channels?
— What alternative channels could be utilized?
— What changes in channel systems can be expected in the future?
— Is the overall promotional mix in terms of spending and allocation of resources the optimal for current circumstances?
— Is the current advertising policy media planning and execution the most appropriate?
— What is the structure of the sales force?
— What revision in sales force structure could improve future effectiveness?
— Is the company utilizing well-formulated below-line and publicity programmes?

Organization is a facet of operations which must be carefully managed if the marketing group is to function effectively. There should be evidence of the current structure having been carefully planned, not created as need arises, as well as evidence of periodic reviews to ensure that any necessary revisions are introduced on a timely basis. In addition, issues such as clear lines of communication, identified responsibilities, overall morale and individual career development paths should all be very clear to the members of the marketing group.

Execution of the actual audit is in itself a control device. However there should also be evidence of information acquisition, analysis and revisionary decisions of the type discussed earlier in the section on performance assessment and efficiency analysis.

Freezing and irradiation of fish
Glossary of UK fishing gear terms
Handbook of trout and salmon diseases
Handy medical guide for seafarers
How to make and set nets
Inshore fishing: its skills, risks, rewards
Introduction to fishery by-products
The lemon sole
A living from lobsters
Marine fisheries ecosystem
Marine pollution and sea life
The marketing of shellfish
Mending of fishing nets
Modern deep sea trawling gear
Modern fishing gear of the world 1
Modern fishing gear of the world 2
Modern fishing gear of the world 3
More Scottish fishing craft and their work
Multilingual dictionary of fish and fish products
Navigation primer for fishermen
Netting materials for fishing gear
Pair trawling and pair seining – the technology of two boat fishing
Pelagic and semi-pelagic trawling gear
Planning of aquaculture development – an introductory guide
Power transmission and automation for ships and submersibles
Refrigeration on fishing vessels
Salmon and trout farming in Norway
Salmon fisheries of Scotland
Scallops and the diver-fisherman
Seafood fishing for amateur and professional
Seine fishing
Squid jigging from small boats
Stability and trim of fishing vessels
The stern trawler
Study of the sea
Textbook of fish culture: breeding and cultivation of fish
Training fishermen at sea
Trends in fish utilization
Trout farming manual
Tuna: distribution and migration
Tuna fishing with pole and line